This report contains the collective views of an international group of experts and does no[illegible] represent the decisions or the stated [illegible] United Nations Environment Programme[illegible] tional Labour Organisation, or the W[illegible] Organization

Environmental Health Criteria 77

MAN-MADE MINERAL FIBRES

Published under the joint sponsorship of
the United Nations Environment Programme,
the International Labour Organisation,
and the World Health Organization

World Health Organization
Geneva, 1988

The **International Programme on Chemical Safety (IPCS)** is a joint venture of the United Nations Environment Programme, the International Labour Organisation, and the World Health Organization. The main objective of the IPCS is to carry out and disseminate evaluations of the effects of chemicals on human health and the quality of the environment. Supporting activities include the development of epidemiological, experimental laboratory, and risk-assessment methods that could produce internationally comparable results, and the development of manpower in the field of toxicology. Other activities carried out by the IPCS include the development of know-how for coping with chemical accidents, coordination of laboratory testing and epidemiological studies, and promotion of research on the mechanisms of the biological action of chemicals.

ISBN 92 4 154277 2

ISSN 0250-863X
PRINTED IN FINLAND
87/7474 — VAMMALA — 5300

CONTENTS

WHO TASK GROUP ON ENVIRONMENTAL HEALTH CRITERIA FOR MAN-MADE MINERAL FIBRES

Members

Dr B. Bellmann, Fraunhofer Institute for Toxicology and Aerosol Research, Hanover, Federal Republic of Germany

Dr J.M.G. Davis, Institute of Occupational Medicine, Edinburgh, United Kingdom[a]

Dr J. Dodgson, Environmental Branch, Institute of Occupational Medicine, Edinburgh, United Kingdom

Professor L.T. Elovskaya, Institute of Industrial Hygiene and Occupational Diseases, Moscow, USSR (Vice-Chairman)

Professor M.J. Gardner, Medical Research Council, Environmental Epidemiology Unit, Southampton General Hospital, Southampton, United Kingdom (Chairman)

Dr M. Jacobsen, Institute of Occupational Medicine, Edinburgh, United Kingdom

Professor M. Kido, Pulmonary Division, University of Occupational and Environmental Health, Fukuoka, Kitakyushu, Japan

Dr M. Kuschner, School of Medicine, State University of New York, Stonybrook, New York, USA[a]

Dr K. Linnainmaa, Department of Industrial Hygiene and Toxicology, Institute of Occupational Health, Helsinki, Finland

Dr E.E. McConnell, Toxicology Research and Testing Program, National Institute of Environmental Health Sciences, Research Triangle Park, North Carolina, USA (Rapporteur)

Dr J.C. McDonald, Dust Disease Research Unit, School of Occupational Health, McGill University, Montreal, Quebec, Canada[a]

Dr A. Marconi, Laboratory of Environmental Hygiene, High Institute of Health, Rome - Nomentana, Italy

Mr I. Ohberg, Rockwool AB, Skövde, Sweden

Dr F. Pott, Medical Insitute for Environmental Hygiene of the University of Dusseldorf, Dusseldorf, Federal Republic of Germany[a]

Dr T. Schneider, Department of Occupational Hygiene, Danish National Institute of Occupational Health, Hellerup, Denmark

Dr J.C. Wagner, Medical Research Council, Llandough Hospital, Penarth, United Kingdom[a]

---

[a] Invited but unable to attend.

Representatives from Other Organizations

Dr A. Berlin, Health and Safety Directorate, Commission of the European Communities, Luxembourg[b]
Ms E. Krug, Health and Safety Directorate, Commission of the European Communities, Luxembourg
Dr R. Murray (International Commission on Occupational Health), London School of Hygiene, London, United Kingdom[b]

Observers

Dr R. Anderson (Thermal Insulation Manufacturers Association), Manville Corporation, Denver, Colorado
Dr D.M. Bernstein, Geneva Facility, Research & Consulting Company AG, Geneva, Switzerland
Dr J.W. Hill (Joint European Medical Research Board), Pilkington Insulation, Ltd., St Helens, Burton-in-Kendal, Cumbria, United Kingdom
Dr O. Kamstrup (Joint European Medical Research Board), Rockwool A/S, Hedehusene, Denmark
Dr J.L. Konzen (Thermal Insulation Manufacturers Association), Medical and Health Affairs, Owens-Corning Fiberglass Corporation, Toledo, Ohio, USA
Dr W.L. Pearson (Canadian Man-Made Mineral Fibre Industry), Fiberglas Canada Inc., Willowdale, Ontario, Canada
Dr G.H. Pigott (European Chemical Industry Ecology and Toxicology Centre), ICI Central Toxicology Laboratory, Alderley Park, Macclesfield, Cheshire, United Kingdom

Secretariat

Dr F. Valic, IPCS Consultant, World Health Organization, Geneva, Switzerland (Secretary)[a]
Ms B. Goelzer, Office of Occupational Health, World Health Organization, Geneva, Switzerland
Dr M. Greenberg, Department of Health and Social Security, London, United Kingdom
Ms M.E. Meek, Bureau of Chemical Hazards, Environmental Health Centre, Health Protection Branch, Health and Welfare Canada, Tunney's Pasture, Ottawa, Ontario, Canada (Temporary Adviser)
Dr L. Simonato, Unit of Analytical Epidemiology, International Agency for Research on Cancer, Lyons, France

---

[a] Head, Department of Occupational Health, Andrija Stampar School of Public Health, Zagreb, Yugoslavia.
[b] Present for part of the meeting only.

NOTE TO READERS OF THE CRITERIA DOCUMENTS

Every effort has been made to present information in the criteria documents as accurately as possible without unduly delaying their publication. In the interest of all users of the environmental health criteria documents, readers are kindly requested to communicate any errors that may have occurred to the Manager of the International Programme on Chemical Safety, World Health Organization, Geneva, Switzerland, in order that they may be included in corrigenda, which will appear in subsequent volumes.

* * *

ENVIRONMENTAL HEALTH CRITERIA FOR MAN-MADE MINERAL FIBRES

A WHO Task Group on Environmental Health Criteria for Man-made Mineral Fibres met at the Monitoring and Assessment Research Centre, London, on 14-18 September 1987.

Dr B. Bennett, who opened the meeting, welcomed the participants on behalf of the host institution. Dr M. Greenberg greeted the participants on behalf the government, and Dr F. Valic welcomed them on behalf of the heads of the three IPCS co-sponsoring organizations (UNEP/ILO/WHO). The Task Group reviewed and revised the draft criteria document and made an evaluation of the health risks of exposure to man-made mineral fibres.

The drafts of this document were prepared by MS M.E. MEEK of the Environmental Health Directorate, Health and Welfare Canada, Ottawa.

The efforts of all who helped in the preparation and finalization of the document are gratefully acknowledged.

* * *

Partial financial support for the publication of this criteria document was kindly provided by the United States Department of Health and Human Services, through a contract from the National Institute of Environmental Health Sciences, Research Triangle Park, North Carolina, USA - a WHO Collaborating Centre for Environmental Health Effects.

1. SUMMARY

1.1 Identity, Terminology, Physical and Chemical Properties, Analytical Methods

Man-made mineral fibres (MMMF), most of which are referred to as man-made vitreous fibres (MMVF), are amorphous silicates manufactured from glass, rock, or other minerals, which can be classified into four broad groups: continuous filament, insulation wool (including rock/slag wool, and glass wool), refractory (including ceramic fibre), and special purpose fibres. Nominal fibre diameters for these groups are 6 - 15 μm (continuous filament), 2 - 9 μm (insulation wools, excluding refractory fibre), 1.2 - 3.5 μm (refractory fibre), and 0.1 - 3 μm (special purpose fibres), respectively. Continuous filament and special purpose fibres are made exclusively from glass, whereas insulation wools can also be made from rock or slag (rock wool or slag wool, also referred to as mineral wool in the USA). Refractory fibres are a large group of amorphous or crystalline synthetic mineral fibres that are highly resistant to heat. They are produced from kaolin clay, from the oxides of aluminium, silicon, or other metals, or, less commonly, from non-oxide materials, such as silicon carbide or nitride. MMMF usually contain a binder and mineral oil as a dust suppressant.

MMMF do not split longitudinally into fibrils of smaller diameter, but may break transversely into shorter segments. The chemical composition of the various MMMF largely determines their chemical resistance and solubility in various solutions, whereas thermal conductivity is also determined by fibre diameter, finer fibres giving lower thermal conductivity.

Analyses for MMMF have been restricted largely to the measurement of total airborne mass concentrations or, more recently (since the early 1970s), to the determination of airborne fibre levels by phase contrast optical microscopy (PCOM). A WHO reference method for monitoring fibre levels by PCOM has been used fairly widely since the early 1980s. Scanning and transmission electron microscopy offer improved resolution and fibre identification in the determination of MMMF. WHO has also developed reference methods to compare and standardize assessments of MMMF by scanning electron microscopy, but the use of these techniques needs to be extended internationally.

1.2 Sources of Human and Environmental Exposure

The global production of man-made mineral fibres has been estimated to be 4.5 million tonnes in 1973 and 6 million tonnes in 1985. Fibrous glass accounts for approximately 80% of MMMF production in the USA, 80% of which is glass wool, mainly used in acoustic or thermal insulation. Textile grades (5 - 10% of fibrous glass production) are used principally for the reinforcement of resinous materials and in textiles, such as draperies. Less than 1% of the production of glass fibre is in the form of fine fibres used in speciality applications, such as high efficiency filter paper and insulation for aircraft. Mineral wool (rock wool/slag wool), which accounts for approximately 10 - 15% of MMMF production in the USA, is used mainly in acoustic and thermal insulation. In Europe, glass wool and rock wool are produced in approximately equal volumes and are also used for thermal and acoustic insulation. Refractory fibres (1 - 2% of all MMMF) are used for high-temperature applications.

There are few quantitative data on emissions of MMMF from manufacturing facilities. Fibre levels in emissions from fibrous glass plants have been reported to be of the order of 0.01 fibres/cm^3. Although quantitative data are not available, it is likely that the emission of MMMF following the installation or disturbance of insulation is the main source of exposure of the general population.

1.3 Environmental Transport, Distribution, and Transformation

Because of the lack of data, only general conclusions on the transport, distribution, and transformation of MMMF in the general environment can be drawn, based on consideration of their physical and chemical properties and of related information concerning the behaviour of natural mineral fibres in ambient air and water. MMMF in ambient air are, on average, shorter and thinner than those in the occupational environment (because of sedimentation of larger diameter fibres and also transverse breakage). In general, most MMMF are more water soluble than naturally occurring asbestiform minerals and are likely to be less persistent in water supplies. They are removed from air by mechanical forces, sedimentation, or thermal destruction, and from water by dissolution and deposition in sediments.

1.4 Environmental Concentrations and Human Exposure

As a general rule, levels of MMMF in the occupational environment have been determined by phase contrast optical microscopy. Average concentrations during the current manufacture of fibrous glass insulation range from 0.01 to 0.05 fibres/cm^3. Levels in continuous fibre plants are an order of magnitude lower, and concentrations in mineral wool plants, in the USA, range up to an order of magnitude higher

(0.032 - 0.72 fibres/cm^3) than those in glass wool production. Under similar plant conditions, mean levels of ceramic fibres are about 4 times higher than those for mineral wool fibres (0.0082 - 7.6 fibres/cm^3). Average concentrations in speciality fine-fibre plants range from 1 to 2 fibres/cm^3 and levels are highest in microfibre production facilities (1 - 50 fibres/cm^3).

In Europe, exposure in MMMF production can be divided historical into early, intermediate, and late technological phases. Whereas the order of magnitude of fibre concentrations in glass wool plants in the early phase is estimated to have been similar to that of current concentrations, concentrations in rock wool and slag wool plants during the early technological phase could have been one or two orders of magnitude higher than those in the late phase. The presence of other contaminants, such as polycyclic aromatic hydrocarbons, arsenic, and asbestos, in the early slag wool production industry has also been reported.

In general, airborne fibre concentrations during the installation of products containing MMMF are comparable to, or less than, levels found in production (< 1 fibre/ml). Exceptions occur during blowing or spraying operations conducted in confined spaces, such as during the insulation of aircraft or attics, when mean levels of fibrous glass and mineral wool have ranged up to 1.8 fibres/cm^3 and 4.2 fibres/cm^3, respectively. Mean concentrations during installation of loose fill in confined spaces have ranged up to 8.2 fibres/cm^3.

Some data are available on the levels of MMMF in ambient and indoor air. Concentrations ranging from 4×10^{-5} to 1.7×10^{-3} fibres/cm^3, measured by TEM, have been found in ambient air. Mean values ranging from 5×10^{-5} to, typically, around 10^{-4} fibres/cm^3, but occasionally over 10^{-3} fibres/cm^3, occur in the air of public buildings.

1.5 Deposition, Clearance, Retention, Durability, and Translocation

Alveolar deposition of MMMF is governed principally by size. For fibres of constant diameter, alveolar deposition decreases with increasing length. On the basis of available data, it has been suggested that there is a rapid decrease in the respirability of fibres > 1 μm in diameter in the rat compared with a suggested upper limit in man of ~ 3.5 μm (5 μm by mouth breathing).

Short MMMF (< 5 μm in length) are efficiently cleared by alveolar macrophages, whereas this form of clearance appears to be much less effective for fibres greater than about 10 μm in length. At present, it is difficult to draw definite conclusions concerning the relative durability of man-made and naturally-occurring mineral fibres *in vivo*. In addition, there are fibres within each of these classes of MMMF that may behave differently from the class as a whole. For instance, longer fibres and fibres of fine diameter dissolve more rapidly in lung tissue than shorter or coarse ones of the same type.

The few available data indicate that translocation of the fibres to other organs and tissues is limited. Fibre concentrations in the tracheobronchial lymph nodes of rats exposed to glass wool and rock wool for 1 year were low, compared with those of glass microfibres. Levels of all fibre types in the diaphragm were essentially zero. The transmigration of fibres appears to be influenced by fibre size and durability, short fibres being present in higher numbers in other tissues than long ones.

1.6 Effects on Experimental Animals and *In Vitro* Test Systems

In the majority of the inhalation studies conducted to date, there has been little or no evidence of fibrosis of the lungs in a range of animal species exposed to glass fibre concentrations of various types of MMMF up to 100 mg/m^3, for periods ranging from 2 days to 24 months. In most studies, the tissue response was confined to accumulation of pulmonary macrophages, many of which contained the fibres. In all cases, the severity of the tissue reaction in animals exposed to glass fibre and, in one study, glass wool, was much less than that for equal masses of chrysotile or crocidolite asbestos. Moreover, in contrast to asbestos, fibrosis did not progress following cessation of exposure. However, the number of asbestos fibres reaching the lung may have been greater than those for fibrous glass and glass wool.

A statistically significant increase in lung tumours has not been found in animals exposed to glass fibres (including glass microfibres) or rock wool in inhalation studies conducted to date. However, in several of the relevant studies, an increase in lung tumours that was not statistically significant was found in exposed animals. In all of the carcinogenicity bioassays conducted to date, similar mass concentrations of chrysotile asbestos have clearly induced lung tumours, whereas crocidolite asbestos has induced few or no tumours. However, available data are insufficient to draw conclusions concerning the relative potency of various fibres types, because the true exposure (number of respirable fibres) was not characterized in most of these studies.

Inhalation or intrapleural injection of aluminium oxide refractory fibre containing about 4% silica caused minimal pulmonary reactions in rats and no pulmonary neoplasms were induced. On the other hand, the incidence of interstitial fibrosis and pulmonary neoplasms following the inhalation of fibrous ceramic aluminium silicate glass was similar to that for chrysotile-exposed animals; however, half of the induced tumours were not typical of those observed in animals exposed to asbestos.

There has been some evidence of fibrosis in various species, following intratracheal administration of glass fibres. However, in most cases, the tissue response has been confined to an inflammatory reaction. An increased incidence of lung tumours has been reported following intratracheal administration of glass microfibres to 2 species in the same laboratory, but these results have not been confirmed by other investigators.

Studies involving intrapleural or intraperitoneal administration of MMMF to animals have provided information on the importance of fibre size and *in vivo* durability in the induction of fibrosis and neoplasia. The probability of the development of mesotheliomas following intrapleural and intraperitoneal administration of these dusts was best correlated with the number of fibres with diameters of less than 0.25 µm and lengths greater than 8 µm; however, probabilities were also relatively high for fibres with diameters of less than 1.5 µm and lengths greater than 4 µm. A model in which the carcinogenic potency of fibres is considered to be a continuous function of length and diameter and also of stability has been proposed. Asbestos has been more potent than equal masses of glass fibre in inducing tumours following intrapleural administration. However, certain types of ceramic fibres were as potent as equal masses of crocidolite asbestos in inducing mesotheliomas after intraperitoneal injection. A similar tumour response was observed after intraperitoneal injection of a comparable number of actinolite asbestos fibres longer than 5 µm, basalt wool, and ceramic wool. Again, individual fibre characteristics are the important criteria for studies using this route of exposure.

In *in vitro* assays, cytotoxicity and cell transformation have also been a function of fibre size distribution, long (generally > 10 µm), narrow (generally < 1 µm) fibres being the most toxic. In general, "coarse" (> 5 µm diameter) fibrous glass (e.g., JM 110) has been less cytotoxic in most assays than chrysotile or crocidolite asbestos. The cytotoxicity of a single type of ceramic fibre was also low. However, the cytotoxicity or transforming potential of fine glass (e.g., JM 100) has approached that of these types of asbestos. With respect to genotoxicity, glass fibres have not induced point mutations in bacterial assays. Glass fibres have been reported to induce delayed mitosis, numerical and structural chromosomal alterations, but not sister chromatid exchanges in mammalian cells *in vitro*. Only a few *in vitro* studies on MMMF other than glass fibres are available.

The effects of fibre coating on the toxicity of MMMF have been examined, to a limited extent, in inhalation and *in vitro* studies. However, the available data are both limited and contradictory and no firm conclusions can be drawn at this time. Effects of combined exposure to MMMF and other pollutants have been examined in inhalation and intraperitoneal injection studies. Concomitant exposure to airborne glass fibres enhanced the toxic effects of styrene in mice, and the incidence of lung cancer in rats exposed by inhalation to radon was increased by concomitant intrapleural injection of glass fibres. In contrast, the carcinogenic potency of glass fibres following intraperitoneal administration was reduced by pre-treatment with hydrochloric acid (HCl).

1.7 Effects on Man

Fibrous glass and rock wool fibres (mainly those greater than 4.5 - 5 μm in diameter) cause mechanical irritation of the skin, characterized by a fine, punctate, itching erythema, which often disappears with continued exposure. However, few reliable data are available concerning the prevalence of dermatitis in workers exposed to MMMF. In several early case reports and in a more recent limited cross-sectional study, eye irritation was also associated with exposure to MMMF in the work-place.

In reports that appeared in the early literature, several cases of acute irritation of the upper respiratory tract and more serious pulmonary diseases, such as bronchiectasis, pneumonia, chronic bronchitis, and asthma, were attributed to occupational exposure to MMMF. However, it is likely that exposure to MMMF was incidental rather than causal in most of these cases, since the reported conditions have not been observed consistently in more recently conducted epidemiological studies.

Some cross-sectional epidemiological studies suggest that there may be MMMF exposure-related effects on respiratory function; others do not. In a large, well-conducted study, there was an increase in the prevalence of low profusion small shadowing on the chest radiographs of cigarette smokers with increasing length of employment in MMMF manufacturing. However, no consistent pattern of MMMF-related non-malignant effects on the respiratory system has emerged, to date, from cross-sectional surveys.

There has been little evidence of excess mortality from non-malignant respiratory disease (NMRD) in MMMF workers in analytical epidemiological studies that have been conducted to date, including the two largest investigations conducted in Europe and the USA. There were no statistically significant increases in NMRD mortality in any sector of the industry in comparison with local rates in the US study, though a statistically significant excess was reported for glass wool workers in comparison with national rates. In the European study, there were no excesses of NMRD mortality. The mortality rates were not related either to time since first exposure or to duration or intensity of exposure.

There has not been any evidence, in studies conducted to date, that pleural or peritoneal mesotheliomas are associated with occupational exposure to MMMF.

An excess of mortality due to lung cancer has been observed in the large epidemiological studies on rock wool/slag wool production workers conducted in Europe and the USA, but not in studies on glass wool or continuous filament workers. The excess of lung cancer mortality and/or incidence in the rock wool/slag wool production industry was apparent when either local or national rates were used for comparison (both statistically significant in the US study and not statistically significant in the European study). There was a relationship (not statistically significant) with time from first exposure, in the European study, but not in the US study. No relationships with duration of employment or estimated cumulative exposure to fibres were observed. In the European study, a statistically significant excess of lung cancer was found in workers in the "early technological phase", during which airborne fibre levels were estimated to have been higher than in later production phases. A statistically significant increase in lung cancer mortality in the European study, 20 years after first exposure, appeared to be associated with the use of slag, but there was a large overlap between the use of slag and the early technological phase. Neither the use of bitumen and pitch nor the presence of asbestos in some products accounted for the observed lung cancer excess. In the European study, there was no excess lung cancer mortality in rock wool/slag wool production workers employed in the "late technological phase", when concentrations of fibres were thought to be lower after the full introduction of dust suppressing agents.

For glass wool production workers, there were no excesses of lung cancer mortality compared with local rates in either the large European or US cohorts, but there were statistically significant increases compared with national rates. In both investigations, mortality from respiratory cancer showed an increase with time from first exposure that was not statistically significant. However, it was not related to duration of employment or estimated cumulative fibre exposure in the US study, or to different technological phases in the European study. The SMR for respiratory cancer, in workers who had been exposed in the manufacture of small diameter (< 3 μm) glass wool fibres in the US cohort, was elevated compared with that in those who had never been exposed in this production sector. The excess in these workers was related to time from first exposure, but neither the overall increase nor the time trends were statistically significant. A statistically significant large excess of lung cancer, observed in a smaller Canadian cohort of glass wool production workers, was not related to time since first exposure or duration of employment.

There has not been an increase in lung cancer mortality or incidence in continuous filament production workers in studies conducted to date.

There have been some suggestions of excesses of cancer at sites other than the lung (e.g., pharynx and buccal cavity, larynx, and bladder) in studies on MMMF production workers. However, in most cases, excesses of cancer of these sites, which were not observed consistently, were small, and were not related to the time since first exposure or duration of employment. Moreover, confounding factors in the etiology of some of the cancers (e.g., alcohol consumption) have not been taken into account in studies conducted to date.

No epidemiological data are available on cancer mortality or incidence in refractory fibre workers.

There have been isolated case reports of respiratory symptoms and dermatitis associated with exposure to MMMF in the home and office environments. However, available epidemiological data are insufficient to draw conclusions in this regard.

1.8 Evaluation of Human Health Risks

1.8.1 Occupationally exposed populations

Data available are insufficient to derive an exposure-response relationship for dermatitis and eye irritation in workers exposed to MMMF. In addition, although there has been some evidence of non-neoplastic respiratory effects in MMMF-exposed workers, no consistent pattern has emerged and it is not possible, therefore, to draw conclusions concerning the nature and extent of the hazard in this regard.

Although there is no evidence that pleural or peritoneal mesotheliomas have been associated with occupational exposure in the production of various MMMF, there have been indications of increases in lung cancer from the principal epidemiological studies, mainly in workers in the rock wool/slag wool sector employed in an early production phase. Although it is possible that other factors may have contributed to this excess, possible contaminants and potential confounding factors examined to date have not explained the excess lung cancer rate. Moreover, available data are consistent with the hypothesis that it is the airborne fibre concentrations that are the most important determinant of lung cancer risk.

In general, airborne MMMF fibre concentrations present in work-places with modern control technology are low. However, mean elevated levels within the same order as those estimated to have been present in the early production phase have been measured in several segments of the production and application industry (e.g., ceramic fibre and small diameter glass wool production and spraying of insulation wool in confined places). The lung cancer risk for the small number of workers employed in these sectors could potentially be elevated if protective equipment is not used.

1.8.2 General population

On the basis of available data, it is not possible to estimate quantitatively the risks associated with exposure of the general population to MMMF in the environment. However, levels of MMMF in the typical general and indoor environments measured to date are much lower (by some orders of magnitude) than some occupational exposures in the past associated with raised lung cancer risks. Thus, the overall picture indicates that the possible risk for the general population is very low, if there is any at all, and should not be a cause for any concern if current low exposures continue.

2. IDENTITY, PHYSICAL AND CHEMICAL PROPERTIES, ANALYTICAL METHODS

2.1 Identity, Terminology, Physical and Chemical Properties

The man-made mineral fibres (MMMF), discussed in this document, will be restricted largely to a subset known as man-made vitreous fibres (MMVF), which are fibres manufactured from glass, natural rock, or other minerals. They are classified according to their source material. Slag wool, rock wool, and glass wool or filaments are produced from slag, natural rock, and glass, respectively (Ottery et al., 1984). Refractory fibres, including ceramic fibres, which are also discussed in this report, comprise a large group of amorphous or partially crystalline synthetic mineral fibres that are highly refractory. They are produced from kaolin clay or the oxides of alumina, silicon, or other metals, or, less commonly, from non-oxide materials, such as silicon carbide, silicon nitride, or boron nitride.

In North America, slag wool and rock wool are often referred to as "mineral wool"; in Europe and Asia, the term "mineral wool" also includes glass wool. A "wool" is an entangled mass of fibres in contrast to the more ordered fibres in continuous filament (textile) glass.

While naturally occurring fibres are crystalline in structure, most man-made mineral fibres are amorphous silicates.[a] The amorphous networks of MMMF are composed of oxides of silica, boron, and aluminium, oxides of the alkaline earth and alkali metals, oxides of bivalent iron and manganese, or amphoteric oxides (e.g., Al_2O_3, Fe_2O_3) (Klingholz, 1977). The chemical compositions of various types of MMMF are presented in Fig. 1 and Tables 1 and 2. Common trade names are given in Table 3 and codes for fibres commonly used in studies of biological effects are included in Table 4. The terminology currently used is not consistent or well defined. Categories and materials are named with reference to use, structure, raw material, or process. The distinctions are not always clear, and categories may overlap.

MMMF products usually contain a binder and an oil for dust suppression. Textile fibre may contain a sizing agent for lubrication. Some special fibres contain a surfactant for improving dispersion. MMMF may be produced with only an oil as a dust suppressant. Special purpose fibres may be produced

---

[a] Most ceramic (aluminosilicate) fibres have an amorphous structure when they are produced, but some conversion to crystalline material (cristobalite, mullite) can occur at high temperature (> 1000 °C) (Vine et al., 1983; Gantner, 1986; Khorami et al., 1986; Strübel et al., 1986).

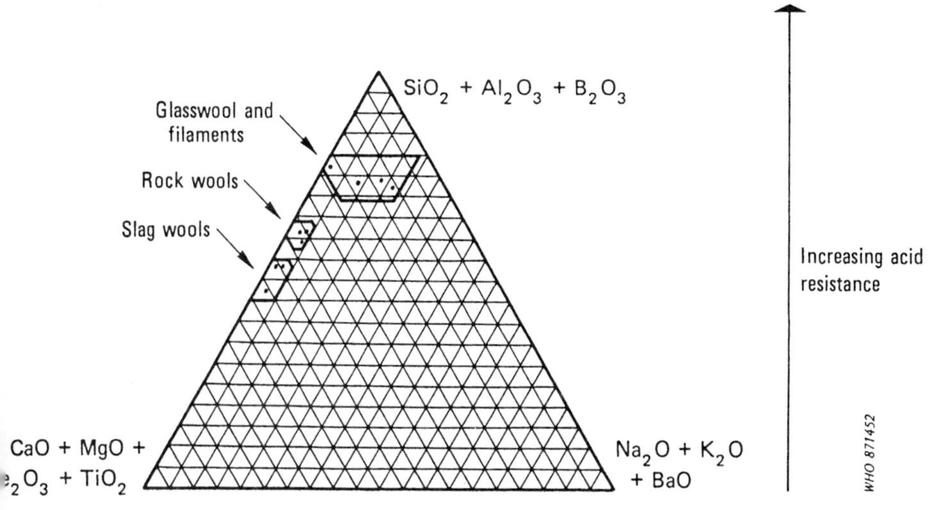

Fig. 1. Chemical composition of man-made mineral fibres. Modified from: Klingholz (1977).

Table 1. Examples of the composition of continuous glass filament, glass wool, rock wool, and slag wool (% by weight)

Component	Continuous glass filament[a]	Glass wool[b]	Rock wool[c]	Slag wool[c]
SiO_2	52 - 56	63	47 - 53	40 - 45
CaO	16 - 25	7	16 - 30	10 - 38
Al_2O_3	12 - 16	(+Fe_2O_3) 6	6 - 13	11.5 - 13.5
B_2O_3	8 - 13	6	-	-
MgO	0 - 6	3	-	-
Na_2O	0 - 3	14	2.3 - 2.5	1.4 - 2.5
K_2O	0 - 3	1	1 - 1.6	0.3 - 1.4
TiO_2	0 - 0.4	-	0.5 - 1.5	0.4 - 2
Fe_2O_3	0.05 - 0.4	-	0.5 - 1.5	8.2
F_2	-	0.7	-	-

[a] From: Klingholz (1977).
[b] From: Mohr & Rowe (1978).
[c] From: IARC (in press).

Table 2. Composition of some commercial ceramic fibres[a] (% by weight)

Component	Fiberfrax[R] bulk	Fiberfrax[R] long staple	Fibermax[Tm] bulk	Fiberfrax[R] HSA	Alumina bulk (SAFFIL[R])	Zirconia bulk	Fireline ceramic	Nextel[R] ceramic fibre 312
Al_2O_3	49.2	44	72	43.4	95	-	95-97.25	62
SiO_2	50.5	51	27	53.9	5	< 0.3	95-97.25	24
ZrO_2	-	5	0	0	0	92	-	-
Fe_2O_3	0.06	0	0.02	0.8	-	-	0.97-0.53	-
TiO_2	0.02	0	0.001	1.6	-	-	1.27-0.70	-
K_2O	0.03	-	-	0.1	-	-	-	-
Na_2O	0.20	-	0.10	0.1	-	-	0.15-0.08	-
CaO	-	-	0.05	-	-	-	0.07-0.04	-
MgO	-	-	0.05	-	-	-	trace	-
Y_2O_3	-	-	-	-	-	8	-	-
B_2O_3	-	-	-	-	-	-	0.06-0.03	14
Leachable chlorides	< 10 ppm	< 10 ppm	11 ppm	< 10 ppm	-	-	-	-
Organics	-	-	-	-	-	-	2.47-1.36	-

[a] From: IARC (in press).

without any additives (Hill, 1977). Some chemicals used in binders are presented in Table 5.

Unlike some natural fibres, MMMF do not split longitudinally into fibrils of smaller diameter, though they may break transversely into shorter fragments. Consequently, the diameters of fibres to which workers may be exposed will depend on the diameter as manufactured, but the length of such fibres will vary according to the extent to which they have been broken subsequently (HSC, 1979).

A particle of any shape, density, and size has a defined settling speed in air. The diameter of a sphere with density 1 g/cm^3 and having the same settling speed as the particle in question will be its aerodynamic diameter (D_{ae}). For MMMF, the D_{ae} is roughly 3 times the geometric diameter, with only a small dependence on fibre length.

2.2 Production Methods

2.2.1 General

The methods of production of various MMMF are presented in Fig. 2. Man-made mineral fibres are produced from a liquid melt of the starting material (e.g., slag, natural rock, glass, clays) at temperatures of 1000 - 1500 °C. Three basic fiberizing methods, which vary from plant to plant and which have changed with technological advances, are used: mechanical drawing, blowing with hot gases, and centrifuging. Combinations

Table 3. Synonyms and trade names of MMMF products[a]

Term/Name	Category	Remarks
TEL	GW	
Fibreglass insulation	GW	Fiberglas[R] is a trade name
Boron silicate glass fibre	GW	Most glass wools are of borosilicate type
Saint Gobain	GW	Major insulation producer (TEL process)
GF/D Whatman filter	GF	Filters made from glass fibres
GF/C microfilter	GF	Filters made from glass fibres
Rock wool	RW	Rockwool[R] is a trade name
Pyrex glass fibres	GF	Pyrex[R] is a glass with high chemical resistance
Basalt wool	RW	
Mineral wool		
Man-made mineral insulation fibres		Slag or rock wool (USA) Glass, slag, or rock wool (Europe)
Insulation wool		
Refractory fibres	CF	
Fibrous ceramic aluminium silicate glass	CF	
Saffil[R]		alumina oxide (4% silica)[b]
Fiberfrax[R]	CF	
Ceramic wool	CF	
Owens-Corning Beta[R]	GF	
Calcium silicate	CF	
Calcium-alumino-silicate	CF	
Refractory ceramic fibre	CF	
Alumina and zirconia fibre	CF	
Zirconia	CF	
Fireline ceramic	CF	
Nextel[R] ceramic fibre	CF	
Fibermak[TM]		

[a] Only synonyms and trade names used in this document are listed.
[b] Microcrystalline.

GW = glass wool.
GF = glass fibre other than wool.
RW = rock wool.
SW = slag wool.
CF = ceramic fibre.

of these methods (e.g., drawing/blowing, blowing/blowing, centrifuging/blowing) are sometimes used (Klingholz, 1977).

MMMF are manufactured to nominal diameters[a] and fall into four broad groups (Fig. 2): continuous filaments, insulation

---

[a] The nominal diameter is the median length-weighted diameter. All fibres in the sample are joined together in order of increasing diameter; the diameter halfway along this long fibre is then the nominal diameter.

Table 4. Codes for Manville glass fibres mentioned in this document[a]

	Designation	Range of nominal diameters (μm)	Glass type[b]
Code:	JM 80	.24 - .28	475
	JM 100	.28 - .38	475
	JM 102	.35 - .42	475
	JM 104	.43 - .53	475, E
	JM 106	.54 - .68	475, E
	JM 110	1.9 - 3	475

a Current data. Specifications have changed over time.
b 475: General purpose borosilicate.
E: Electrical grade, alkali-free borosilicate.

Table 5. Chemicals used in binders for MMMF[a]

Phenol formaldehyde resin
Urea formaldehyde resin
Melamine formaldehyde resin
Polyvinyl acetate
Vinsol resin
Urea
Silicones
Dyes
Ammonium sulfate
Ammonium hydroxide
Starch
Carbon pigment
Epoxy resins
Pseudo-epoxy resins
Bitumens

a From: Hill (1977) and WHO (1983a).

wools, refractory fibres, and special purpose fibres (Hill, 1977; Ottery et al., 1984). Continuous filaments, used in textiles and for reinforcing plastics, are produced by the drawing process and individual fibre diameters (6 - 15 μm) are closely distributed around the median value. Few are respirable. The nominal diameter of insulation wools, which are produced largely by centrifuging or blowing or a combination of both, is about 4 or 5 μm.[a] The production method gives a much wider distribu-

a Some glass fibre insulation wool with a nominal fibre diameter of about 2 μm is now being produced in North America.

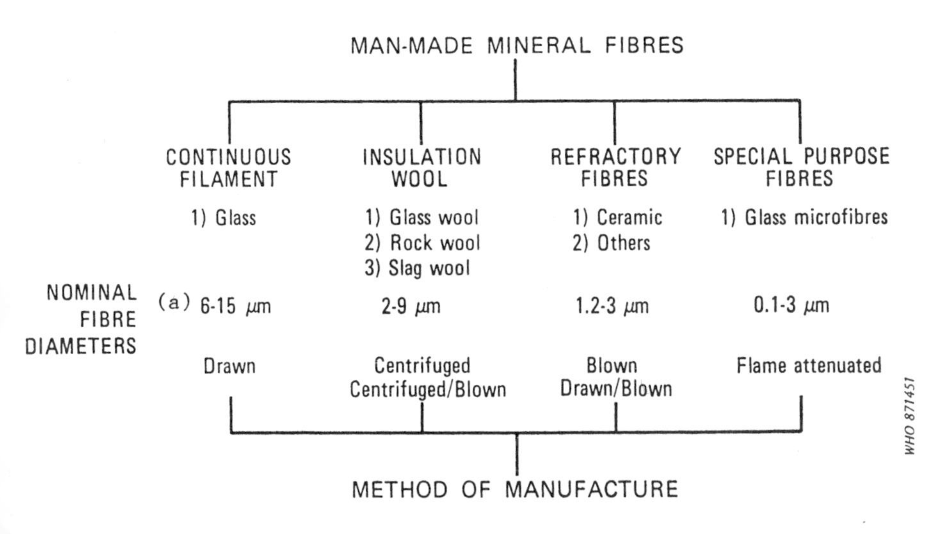

Fig. 2. Classification, methods of manufacture, and nominal diameters of MMMF. Modified from: Head & Wagg (1980).

tion around the nominal diameter, with a high proportion of respirable fibres. For special applications, such as hearing protection, the nominal diameter may be as low as 1 - 2 µm. Special fibres, which account for only about 1% of world production, are used in special applications, such as high-efficiency filter papers (Hill, 1977) and insulation for aircraft and space vehicles (Kilburn, 1982). The diameters of the majority of these fibres are less than 1 µm. Continuous filaments and special purpose submicron-range fibres are made exclusively from glass, whereas insulation wools can also be manufactured from rock or slag.

Available data on the physical and chemical properties of some MMMF are presented in Table 6. The chemical composition of the various MMMF determines their chemical resistance (the sum of the acidic oxides divided by the sum of the amphoteric and basic oxides measured in molar units). Typical values are 0.50 - 0.65 for slag wool, 0.80 - 1.10 for rock wool, and 1.55 - 2.50 for glass wool (Klingholz, 1977). The solubility of MMMF in aqueous and physiological solutions varies considerably, according to their chemical composition and fibre size distribution. Solubility increases with increasing alkali content for a given composition of other elements, and fine fibres degrade more rapidly than coarse ones *in vitro* (Spurny et al., 1983). Thermal conductivity is mainly a function of fibre diameter and bulk density, with finer fibres having a lower thermal conductivity.

Table 6. Physical and chemical properties of some MMMF

MMMF	Fibre size distribution[a] (µm)	Fibre surface area[b] (m^2/g)	Melting point[b,c] (°C)	Density (g/cm^3)	Refractive index[a]	Physical state[b]
Glass filament	range of fibre diameters: 6 - 9.5	-	-	2.596[a]	1.548	
Glass wool (coarse)	range of fibre diameters: 7.5 - 15	-	-	2.605[a]	1.549	
Glass wool	range of fibre diameters:	-	-			
- micro glass fibres	0.75 - 2			2.568[a]	1.537	
- special purpose glass fibres	0.25 - 0.75					
Ceramic (Fiberfrax[R] bulk)	mean fibre diameters: 2 - 3; mean fibre length: < 102 mm	0.5	1790	2.73[b]	-	white fibre
Ceramic (Fiberfrax[R] long staple)	mean fibre diameters: 5 and 13; mean fibre length: < 254 mm	-	1790	2.62[b]	-	white fibre
Ceramic (Fibermax[TM] bulk)	mean fibre diameters: 2 - 3.5	7.65	1870	3[b]	-	white, mullite, polycrystalline

Table 6 (contd).

Ceramic (Fiberfrax[R] HSA)	mean fibre diameters: 1.2; mean fibre length: 3 mm	2.5	1790	2.7[b]	-	white to light grey
Ceramic (SAFFIL[R] - alumina bulk)	mean fibre diameters: 3; mean fibre length: 3 mm	-	2040	3[b]	-	white
Ceramic (zirconian bulk)	mean fibre diameters: 3 - 6; mean fibre length: < 1.5 mm	-	2600	0.24-0.64[b]	-	white
Fireline ceramic	-	-	1700	-	-	white to cream colour
Ceramic (Nextel[R] fibre 312 filament)	mean fibre diameters: 8 - 12; mean fibre length: continuous	< 1	1700	> 2.7[b]	-	white, smooth, transparent, continuous polycrystalline metal oxide

a From: NIOSH (1977).
b From: IARC (in press).
c Vitreous fibres have no precies melting point.

2.2.2 Historical

Production conditions in early mineral wool plants were usually very primitive, particularly those operating before 1950. Many different raw materials were used, especially in the slag wool industry. In addition to foundry and steel slags, chrome, lead, and copper slags were also used. The glass wool industry, being based on an already existing raw material and melting technology, was technologically more advanced and stable (Ohberg, in press).

The production environment during these early operations was poorly controlled and could occasionally be heavily polluted.

Some plants started in a batch operation mode and operated this way for a period that varied between plants. Subsequently, production methods changed to continuous operating procedures (Cherrie & Dodgson, 1986; Ohberg, in press). Oil was not usually used as a dust suppressant during these early operations.

In a historical environmental investigation (Cherrie & Dodgson, 1986), conducted as part of the large European epidemiological study (Simonato et al., 1986a, in press), the absence of oil and or the use of batch production was defined as the early technological phase. The introduction of dust-suppressing agents and implementation of continuous production procedures constituted the late phase. Between the early and late phases, in most production facilities, there was a transition period called the intermediate phase.

Methods of product manufacture and manipulation were principally manual, the products were very simple, and production rates were low in the early phase. Later, the diversity and complexity of the products increased as well as the production rate. The degree of mechanization was also increased (i.e., introduction of saws, cutters, conveyors, etc.), with a corresponding gradual reduction of manual handling (Cherrie et al., in press; Dodgson et al., in press; Ohberg, in press).

2.3 Analytical Methods

2.3.1 Air

The methods currently used for the collection, quantification, and identification of airborne fibrous particulates in the occupational and general environments are summarized in Table 7. The detection limits shown in Table 7 are optimal values. Rather higher detection limits are achieved in routine practice; namely, 0.25 μm for PCOM, 0.05 μm for SEM, and 0.005 μm for TEM.

For MMMF, analyses have been restricted largely to the measurement of total airborne mass concentration or, more

Table 7. Measurement of MMMF

Measurement	Sampling	Quantification	Identification	Detection limit (µm)	Comments	Reference
Mass	glass wool or cellulose ester membrane filters; pore size 0.6 - 1.5 µm	gravimetric (analytical balance)			biologically relevant fraction not determined	WHO (1981)
Fibre number	cellulose ester membrane filters; pore size 0.5 - 1.5 µm	phase contrast optical microscopy (PCOM)	morphology	2×10^{-1}	fibres with diameter < 0.2 µm not detected; not possible to distinguish MMMF from other fibres	WHO (1981, 1985); Chatfield (1983)
Fibre number	membrane or Nuclepore polycarbonate filters	scanning electron microscopy (SEM)	morphology; chemical composition by EDXA[a]	5×10^{-4}	improved resolution and fibre identification	Burdett & Rood (1983); Chatfield (1983)
Fibre number	membrane or Nuclepore filters	transmission electron microscopy (TEM)	morphology; chemical composition by EDXA; crystalline structure by SAED[b]	2×10^{-5}	best resolution and fibre identification; sample preparation not standardized	Chatfield (1983)

[a] EDXA - Energy dispersive X-ray analysis.
[b] SAED - Selected area electron diffraction.

recently, to the determination of airborne fibre number concentrations by phase contrast optical microscopy (PCOM). The method for sampling personal exposure levels involves drawing a measured volume of air through a filter mounted in a holder that is located in the breathing zone of the subject. Static sampling methods are not recommended for measuring personal exposures. When measuring mass concentrations of MMMF, either cellulose ester membrane or glass fibre filters can be used. The filters are stabilized in air and weighed against control filters, both before and after sampling, to permit correction of weight changes caused by varying humidity. Only cellulose ester filters are used for assessing fibre number concentrations. In this case, the filter is made optically transparent with one of several clearing agents (e.g., triacetin, acetone, or ethylene glycol monomethyl ether) and the fibres present within random areas are counted and classified using PCOM. For this purpose, a fibre is defined as a particle having a length to diameter ratio (aspect ratio) $\geq$ 3 and a length $\geq$ 5 μm. Respirable fibres are those with diameters $\leq$ 3 μm. Fibres with diameters exceeding 3 μm are termed non-respirable fibres.

Sampling strategies should be well-designed, based on careful consideration of "how", "where", "when", and "for how long" to sample, as well as "how many samples" to collect to ensure that the results are comparable. Sampling strategies will vary depending on the reason for sampling, e.g., epidemiology, dust control, etc. Sampling strategy has been discussed in the literature (NIOSH, 1977; Valić, 1983; WHO, 1984).

Although the basic methods for the determination of total airborne mass and fibre number concentrations in most countries are similar, differences in the sampling procedure, the filter size and type, the clearing agent, and the microscope used, and, particularly, statistical and subjective errors in counting, contribute to variations in results. In order that results from different countries should be more comparable, a reference method for monitoring MMMF at the work-place based on PCOM was proposed by WHO in 1981 and has been widely used in a slightly modified form (WHO, 1985) since that time. The results of recent slide exchanges among participating laboratories have shown a maximum systematic counting difference of about 1.8 times for this method (Crawford et al., in press).

Rendall & Schoeman (1985) reported that the contrast of the phase image for the counting of glass fibres was improved by the low-temperature ashing of the membrane filter attached to glass slides by acetone vapours followed by microscopic examination using simple Köhler illumination. Automatic counting methods are currently being developed and may in future provide greater consistency in results (Burdett et al., 1984).

The improved resolution of electron microscopy and the identification capacity, particularly of the analytical transmission electron microscope (TEM with selected area electron diffraction (SAED) and energy dispersive X-ray analysis (EDXA)), make these methods more suitable for analysis of fibres in the general environment, where MMMF may constitute only a small fraction of the airborne fibrous material. However, to date, use of these methods for the determination of MMMF has been restricted largely to the characterization of fibre sizes airborne in the occupational environment.

In analysing by SEM, the fibres collected on polycarbonate filters can be directly examined. This avoids the need to use transfer techniques that may affect the fibre size distribution. WHO has developed a reference method for SEM, principally to characterize fibre size (WHO, 1985). Using the same sampling method as for the PCOM method, samples are collected on a polycarbonate (Nuclepore) or PVC-copolymer membrane filter (Gelman DM 800) and observed at a magnification of 5000 times. Fibre lengths and diameters are measured from optically enlarged images of photomicrographs.

In the past, methods of sample preparation for TEM have varied considerably, making comparison of values obtained by different investigators difficult. At present, direct transfer preparation techniques involving carbon coating of particles on the surface of a polycarbonate or membrane filter and indirect sample preparation methods, in which attempts have been made to retain the fibre size distribution, are the most widely accepted for analysis of fibres in air and water (Toft & Meek, 1986).

The size distribution of "superfine" MMMF has been successfully measured by a direct transfer sampling method on a membrane filter, subsequent analysis by analytical TEM, and fibre measurement by an image analysis system (Rood & Streeter, 1985). On the basis of their results, the authors concluded that a substantial proportion of the fibres would not have been detected by PCOM or SEM, even fibres longer than 5 μm.

2.3.2 Biological materials

Several techniques have been developed for the recovery and determination of mineral fibres in biological tissues. Fibres are commonly separated from tissue samples by digestion (e.g., by sodium hypochlorite or potassium hydroxide) or ashing (both low and high temperature) and identified subsequently by light or electron microscopy. Methods for sampling, analysis, and identification of mineral fibres in lung tissues have been reviewed recently by Davis et al. (1986).

To date, methods for the sampling and analysis of tissues for fibrous particles have not been standardized and it is difficult to compare the results of various investigators.

Moreover, available data indicate that common methods of tissue separation, such as tissue digestion with sodium hypochlorite or potassium hydroxide, cause substantial losses of MMMF (Johnson et al., 1984b). Thus, data on levels of MMMF in biological tissues should be cautiously interpreted.

A method for the sampling and determination of fibres in the eyes of workers handling MMMF has been described by Schneider & Stokholm (1981). Mucous threads and dried mucous, stained with alcian blue, were removed from the eye, placed on a slide, ashed in a low-temperature asher, and examined for non-respirable fibre content by optical microscopy. The authors reported a good correlation, among 15 samples, between fibre levels in the eyes and total dust exposure or total non-respirable fibre exposure. Levels were not correlated with airborne respirable fibre concentrations.

3. SOURCES OF HUMAN AND ENVIRONMENTAL EXPOSURE

3.1 Production

Few recent quantitative data are available concerning the global production of MMMF. As illustrated in Table 8, this value was reported to be 4.5 million tonnes in 1973 (WHO, 1983a; Järvholm, 1984). In 1976, the value of the production of MMMF was estimated to be greater than 1 billion US dollars (Corn, 1979).

Table 8. Estimated world production of MMMF materials in 1973[a]

Area	Insulation materials		Textile materials		Total	
	1000 tonnes	%	1000 tonnes	%	1000 tonnes	%
America						
Central/South	120	3	20	2	140	3
North America	1600	43	400	46	2000	43
Australia	30	1	-	-	-	-
Europe						
Eastern	600	16	85	10	685	15
Western	1200	32	260	30	1460	32
Japan	200	5	100	12	300	7
World	3750	100	865	100	4585	100

[a] From: WHO (1983a).

It has been estimated that, in 1985, total world production of MMMF was between 6 and 6.5 million tonnes, of which insulation wools accounted for approximately 5 million tonnes and textile grades for 1 - 1.5 million tonnes (EURIMA, 1987)[a].

Estimated annual production of fibrous glass in the USA in the late 1970s was 370 000 tonnes for textile fibres and 1 200 000 tonnes for glass wool. Total annual production in the USA has been estimated to be 200 000 tonnes for mineral wool (slag and rock wool) and 21 000 tonnes for ceramic fibre (NRC, 1984). In the United Kingdom, production of wool materials rose from 3500 tonnes in 1937 to 130 000 tonnes in 1977 (WHO, 1983a).

[a] Information provided by the European Insulation Manufacturers Association, Luxembourg.

3.2 Uses

The main uses of MMMF are presented in Fig. 3. Fibrous glass accounts for approximately 80% of MMMF production in the USA, with 80% of this being wool fibres used mainly in thermal or acoustic insulation (Kirk-Othmer, 1980). Insulation wools are usually compressed into "bats", "boards", "blankets", "sheets" etc. or bagged as loose wool for blowing or pouring into structural spaces. Some products are sewn or glued on to asphalt paper, aluminium foil, etc. Textile grades, which account for 5 - 10% (or greater) of all fibrous glass, are used extensively in reinforcing resinous materials (e.g., automobile bodies or boat hulls), paper and rubber products, in textiles, such as draperies, and in electrical insulation and cording. Less than 1% of the production of glass fibre is in the form of fine fibres (fibre diameters < 1 μm), which are used in specialty applications such as high efficiency filter papers and insulation for aircraft and space vehicles.

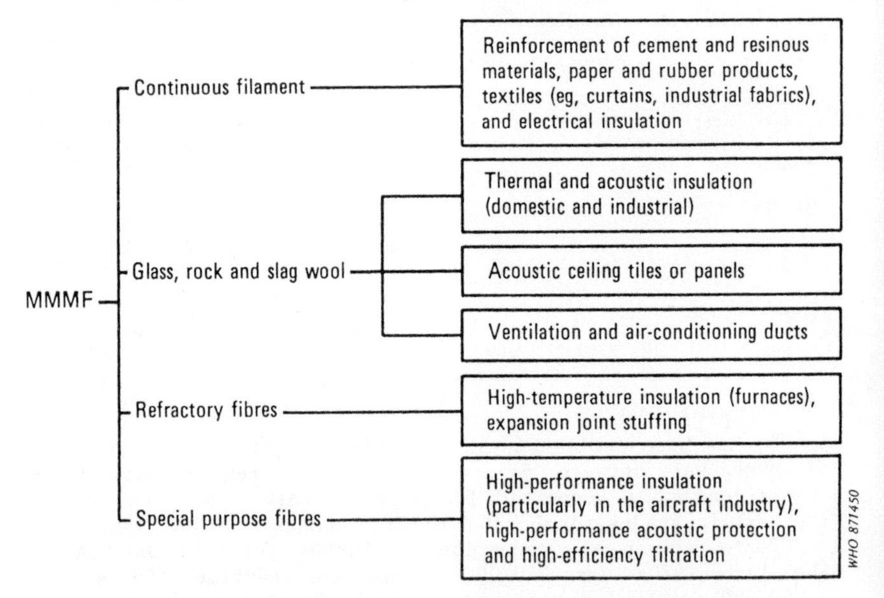

Fig. 3. Uses of MMMF. Prepared on the basis of information included in: Kirk-Othmer (1980), TIMA (1982), WHO (1983a), and NRC (1984).

Rock wool and slag wool, which account for approximately 10 - 15% of MMMF production in the USA, are used mainly as acoustic or thermal insulation for industrial buildings or processes. In Europe, where the production volumes of glass wool and rock wool are similar, rock wool and slag wool are also used extensively for domestic insulation.

Ceramic fibres (1 - 2% of all MMMF) are used mainly for high-temperature insulation, e.g., thermal blankets for industrial furnaces. Smaller quantities are used for expansion joint filling.

3.3 Emissions into the Environment

Data on emissions of MMMF from manufacturing facilities are limited. Fibre levels in emissions from fibrous glass and mineral wool plants in the Federal Republic of Germany, determined by SEM, were of the order of 10^{-2} fibres/cm^3 (Tiesler, 1982). On the basis of these data, it was estimated that the total fibrous dust emissions from MMMF plants in this country were about 1.8 tonnes/year. Emissions of respirable fibres (defined by the investigators as fibres with lengths exceeding 10 μm and diameters of less than 1 μm) were estimated to be 80 kg/year.

It seems likely that the main source of emissions of MMMF (mainly glass fibres) in indoor air is insulation in public buildings or homes. Although quantitative data are not available, emissions are probably highest shortly after installation or following disturbance of the insulation (NRC, 1984).

4. ENVIRONMENTAL TRANSPORT, DISTRIBUTION, AND TRANSFORMATION

The transport, distribution, and transformation of MMMF in the general environment have not been specifically studied. However, some general conclusions can be drawn, on the basis of the physical and chemical properties of MMMF and information concerning the behaviour of natural mineral fibres in the ambient air and water.

Mineral fibres with small diameters are most likely to remain airborne for long periods. Marconi et al. (in press) and Riediger (1984) reported that, during operations involving the use of mineral wool products, there is a general trend for airborne fibres to become shorter and thinner with increasing distance from working areas. For example, during the installation of rock wool blankets, Marconi et al. (in press) reported that the respirable fraction of airborne fibres in the working areas accounts for about 67% of total airborne fibres. At a distance of about 5 m from the working area, the respirable fraction is about 90% of the total fibres. This is attributable to the sedimentation of larger diameter fibres.

The solubility of MMMF in water varies considerably as a function of their chemical composition and fibre size distribution (section 2). Solubility increases with increasing alkali content for a given composition of other elements, and fibres with fine diameters degrade more rapidly than coarse ones (Spurny et al., 1983). However, in general, MMMF are more water soluble than naturally occurring asbestiform minerals and, thus, most are likely to be less persistent in water supplies.

MMMF can be removed from the environment by breakage into successively smaller fragments, thus losing their fibre characteristics, by sedimentation and subsequent burial in soil, or by thermal destruction (e.g., during incineration of MMMF-containing waste). Dissolution and deposition and subsequent burial in sediments are the most likely mechanisms of removal from water.

5. ENVIRONMENTAL CONCENTRATIONS AND HUMAN EXPOSURE

5.1 Environmental Concentrations

5.1.1 Air

5.1.1.1 Occupational environment

(a) Production

Available data on levels of MMMF in production industries are presented in Tables 9 and 10. Although in most of the surveys conducted to date, both mass concentrations of particulate matter and respirable fibre levels have been determined, this discussion will be restricted largely to information on fibre concentrations that appear to be most relevant for the evaluation of potential health effects.

Comparison of data from various surveys is complicated by lack of consistency in the classification of various operations and job categories and by differences in sampling strategy and fibre counting criteria. For example, a reference method for the determination of airborne MMMF in the occupational environment has been introduced only relatively recently (WHO, 1981, 1985); reassessment using the WHO PCOM reference method of levels determined previously by interference microscopy resulted in an increase in reported airborne concentrations in European insulation plants of approximately two-fold (Cherrie et al., 1986). However, in spite of the difficulties in comparing data, some general conclusions concerning airborne fibre size and concentrations in occupational environments can be drawn. Fibre levels in plants manufacturing MMMF are substantially lower than those measured in factories using asbestos. In addition, in general, the measured median diameter of airborne fibres is consistently and substantially smaller than the nominal diameter of the fibres in the product. The median airborne fibre diameters were 4 μm, 1.5 μm, and 0.2 μm for a product of nominal fibre diameters of 14 μm, 6 μm, and 1 μm, respectively (Esmen et al., 1979a). Esmen et al. (1979a) have shown that there is a negative correlation ($r = -0.96$) between the overall process average exposure levels (fibres/cm^3) and the size of the nominal diameter produced. A similar pattern has been observed by Cherrie & Dodgson (1986).

The average concentrations measured by PCOM during the manufacture of fibrous glass insulation are of the order of 0.03 fibres/cm^3. Average concentrations in continuous fibre plants are about one order of magnitude lower and concentrations in mineral wool (rock and slag) plants in the USA range up to one order of magnitude higher. Corresponding concentrations in

Table 9. Airborne concentrations of MMMF in production industries

Sampling	Analysis	Results		Reference
		Occupational group mean fibre concentrations in fibres/cm^3	Fibre size distribution	
After 1970: 650 samples in 6 wool insulation, 5 continuous textile, and 4 fibre glass-reinforced plastic products plants	Phase contrast optical microscopy; length: diameter $\geq$ 3:1		% fibres with diameter < 3.5 µm:	Konzen (1976)
		Glass wool plants, 0.11 - 0.16	68.9 - 87.1	
		Continuous textile, 0.07 - 0.37	76.9 - 98.0	
		Non-corrosive product, 0.12	72.3	
		Flame attenuated fibre production, 0.38	89.3	
1975-78: > 1500 personal 8-h samples in 16 production facilities (loose, continous, and mixed) including 5 mineral wool plants in the USA	Phase contrast optical microscopy; length: diameter $\geq$ 3:1; length > 5 µm; transmission electron microscopy	Estimate of exposure for various operations:	Concentration of fibres with diameter < 1.5 µm and length > 8 µm:	Corn & Sansone (1974); Corn et al. (1976); Esmen et al. (1978, 1979a); Corn (1979); Esmen (1984)
		Continuous, < 0.003	0	
		Coarse, 0.001 - 0.005	10^{-5} - 10^{-4}	
		Fibre glass insulation, 0.01 - 0.05	10^{-3} - 10^{-2}	
		Mineral wool, 0.2 - 2	5×10^{-2} - 5×10^{-1}	
		Specialty-fine, 1 - 2	0.5 - 1	
		Microfibre, 1 - 50	0.5 - 20	

Table 9 (contd).

> 950 breathing zone and static general atmospheric > 4-h samples at 25 manufacturing plants and construction sites in the United Kingdom	phase contrast optical microscopy; length: diameter ≥ 3:1; length > 5 µm; diameter < 3 µm; 10% re-examined by interference microscopy for fibre size distribution	Continous filament glass fibres: mean < 0.02; insulation wools-fibre production: 0.12 - 0.89; production of basic fibrous materials (slab, board, etc): 0.02 - 0.37; conversion of basic materials to finished products (e.g., pipe sections): 0.02 - 0.35; special purpose fibres: 0.8 - 3.70; ceramic fibre-manufacture and conversion: 1.09 - 1.27	% with diameter < 1 µm: glass wool insulation - 16%; mineral wool insulation - 18%; glass microfibres - 60%; ceramic fibre - 18%	HSC (1979); Head & Wagg (1980)
After 1980: > 200 personal 8-h samples in 7 plants (2 mineral wool, 2 ordinary glass fibre, 2 ordinary and fine glass fibre, and 1 very fine glass fibre) in the USA	phase contrast optical microscopy; length: diameter > 3:1; length > 5 µm; transmission electron microscopy	In order of decreasing concentrations: Very fine glass fibre plant: 0.048 - 6.77 Mineral wool plants: 0.032 - 0.72 Ordinary and fine glass fibre plants: 0.014 - 2.22 Ordinary glass fibre plants: 0.017 - 0.062	Tendency for fibre diameters to be smaller than in the first survey	Hammad & Esmen (1984)
> 300 personal and stationary samples in 11 plants (4 large diameter insulation, 6 manufacturing or using small diameter fibres, and 1 making fibrous glass reinforced products) in the USA	phase contrast optical microscopy: all visible fibres; electron microscopy for a proportion of samples	 Large diameter glass insulation plants, 0.04 - 0.20 Small diameter glass fibre plants, 0.8 - 21.9 Fibrous glass reinforced plastics plant, 0.03 - 0.07	% with diameter < 1 µm: 2 - 46% 62 - 96% < 1%	Dement (1975)

Table 9 (contd).

Sampling	Analysis	Results		Reference
		Occupational group mean fibre concentrations in fibres/cm^3	Fibre size distribution	
1981: unspecified number of samples at 75 locations in plants in the Federal Republic of Germany	phase contrast optical microscopy: length: diameter > 3:1; length > 5 µm; diameter < 3 µm; scanning electron microscopy for a proportion of samples (40 out of 75)	Centrifuging/blowing: < 0.1 - 0.44 Centrifuging rock and slag wool: < 0.1 - 0.4 Blowing rock wool: < 0.1 - 0.5 Cutting glass fibre filters: 0.1 - 0.2 Glass fibre reinforced plastics (grinding and cuffing): < 0.1 - 0.4	Less than 1% of fibres with length > 20 µm and diameter < 0.25 µm	Riediger (1984)
Unspecified number of static samples in 3 rock wool and 1 continuous filament plant	phase contrast optical microscopy: length: diameter > 3:1; length > 5 µm; diameter < 3 µm	Rock wool: 0.10 - 0.65 Glass fibre (continuous filament): 0.10 - 0.46	% of fibres with diameter < 1.5 µm: 17 - 53.5 27	Indulski et al. (1984)

Table 9 (contd).

430 personal 8-h samples in 3 ceramic fibre production and product manufacturing plants	phase contrast optical microscopy; length: diameter ≥ 3:1; length > 5 µm; transmission electron microscopy	Mean concentrations ranged from 0.0082 (fibre cleaning) to 7.6 (process helper); individual range, 0.0012 - 56	Geometric mean diameter, 0.7 ± 0.2 um; ~90% < 3 µm in diameter; diameters of airborne fibres from manufacturing operations greater than those from production zones	Esmen et al. (1979b)
10 samples each from 10 rock and glass wool plants in Europe examined by Ottery et al. (1984) plus 20 "high" and "low" density samples reassessed using WHO PCOM[a] reference method	phase contrast optical microscopy; length: diameter ≥ 3:1; length ≥ 5 µm; diameter < 3 µm; scanning electron microscopy	Revised fibre concentrations approximately twice the original levels; mean levels in insulation wool plants < 0.1 and in continuous filament plants < 0.01	Median fibre lengths: rock wool plants, 10 - 20 µm; glass wool plants, 8 - 15 µm; median diameters: rock wool, 1.2 - 2 µm; glass wool, 0.7 - 1 µm; 20 - 50% of fibres with lengths > 8 µm and diameters < 1.5 µm (% higher in glass wool than in rock wool plants)	Cherrie et al. (1986)

[a] PCOM - Phase contrast optical microscopy.

Table 10. Airborne gravimetric concentrations of MMMF in production industries

Sampling	Analysis	Results		Reference
		Short-term samples (mg/m^3)	Time-weighted average concentration (mg/m^3)	
1972; unspecified number of samples in glass fibres plant	Total airborne dust in breathing zone		0.1 - 0.75	Roschin & Azova (1975)
Mullitosilica fibres	Gravimetry, chemistry, and microscopy	2.6 - 10.5 (loading in equipment, cutting, and packing glass fibres) SiO_2:Al_2O_3 ~1.0 SiO2 ~45.2% Al2O3 ~53.1% Fe_2O_3 - 0.09% CaO - 0.19% MgO - 0.34 K_2O - 0.11 Na_2O - 0.19 length: 1 - 3 µm < 2.3% 4 - 6 µm - 4.3 - 7.6% 7 - 11 µm - 7 - 11.3% 12 - 17 µm - 10.5 - 13% 18 - 25 µm - 17.3 - 20.3% > 25 µm - 38 - 48.7% diameter: 1 - 3 µm	2 - 3	Skomarokhin (1985)

European rock wool plants are of the order of 0.1 fibres/cm^3. Under similar plant conditions, airborne fibre concentrations are about 4 times higher in ceramic fibre production than in US mineral wool (rock and slag) plants, and average ceramic fibre concentrations have been reported to range from 0.0082 to 7.6 fibres/cm^3 (Esmen et al., 1979b). Average airborne concentrations in speciality fine fibre plants range from 1 to 2 fibres/cm^3 and concentrations are highest in microfibre production facilities (1 - 50 fibres/cm^3). However, it should be noted that individual exposure varies considerably; for individuals in the same job classification, concentrations range over 2 orders of magnitude (Corn, 1979).

Total dust concentrations are typically of the order of 1 mg/m^3, irrespective of the fibre type manufactured (excluding ceramic). Overall averages were 4 - 5 mg/m^3 for one rock wool and one glass wool plant where manufacturing was reported to be heavy or very heavy (Esmen et al., 1979a). The situation in 13 European plants was similar (Cherrie et al., 1986). Averages for 3 US ceramic fibre production plants were 6, 1.6, and 0.85 mg/m^3, respectively. Results from a USSR ceramic fibre plant are comparable (Skomarokhin, 1985).

With respect to the relationship between fibre and mass concentrations, Ottery et al. (1984) summarized their observations as follows:

"Where fibre and mass concentrations were compared on a plant-average basis a broad correlation was observed (in general, those plants which are "dusty" are also the ones with higher airborne fibre concentrations). However, this relation was not consistent between different occupational groups, nor was there any detectable correlation when mass and fibre concentrations were considered on an individual basis."

This is generally consistent with the observations of other investigators (Esmen et al., 1978; Head & Wagg, 1980).

(b) User industries

Available data on airborne fibre concentrations associated with the installation of products containing MMMF are presented in Table 11. Airborne concentrations during the installation of insulation vary considerably, depending on the method of application and the extent of confinement within the work-space. Concentrations during installation are comparable to, or lower than, levels found in the production of the fibrous material (Esmen et al., 1982), with the exception of blowing or spraying operations conducted in confined spaces, such as during the insulation of aircraft or attics (Head & Wagg, 1980; Esmen et al., 1982). In various surveys, mean concentrations during

Table 11. Airborne concentrations of MMMF in application industries

Application	Number of samples	Analysis	Concentration (fibres/cm^3) Mean (range)	Reference
Denmark				
Attic insulation, existing buildings	total of 200 samples at 24 sites	phase contrast optical microscopy; length: diameter > 3:1; length > 5 µm; diameter < 3 µm; scanning electron microscopy	0.89 (0.04 - 3.5)	Schneider (1979, 1984)
Insulation, new buildings			0.10 (0.04 - 0.17)	
Technical insulation			0.35 (0.03 - 1.6)	
Italy				
Ship insulation (rock wool blankets)				
- inside room	8	phase contrast optical microscopy; length: diameter > 3:1; length > 5 µm; diameter < 3 µm	0.19 (0.05 - 0.65)	Marconi et al. (1986)
- outside room	9		0.021 (0.002 - 0.05)	
- installers	14		0.13 (0.009 - 0.41)	
- "finishing" workers	14		0.12 (0.03 - 0.31)	
Sweden				
Attic insulation, existing buildings	total of 58 samples at 14 sites	phase contrast optical microscopy; length: diameter > 3:1; length > 5 µm; diameter < 3 µm	1.1 (0.1 - 1.9)	Hallin (1981); Schneider (1984)
Insulation, new buildings			0.57 (0.007 - 1.8)	
Technical insulation (boilers, cisterns)			0.37 (0.01 - 1.39)	
Acoustic insulation			0.15 (0.11 - 0.18)	
- spraying			0.51 (0.13 - 1.1)	
- hanging fabric			0.60 (0.30 - 0.76)	

Table 11 (contd).

United Kingdom				
Domestic loft insulation		phase contrast optical microscopy; length: diameter ≥ 3:1; length > 5 µm; diameter < 3 µm		Head & Wagg (1980)
- glass fibre blanket	12		0.70 (0.24 - 1.8)	
- loose fill-mineral wool	16		8.2 (0.54 - 21)	
Fire protection-structural steel				
- sprayed mineral wool	22		0.77 (0.16 - 2.6)	
Application in industrial products				
- industrial engine exhaust insulation (mineral wool)	15		0.085 (0.02 - 0.36)	
USA				
Fibrous glass duct wrapping	5	phase contrast optical microscopy; size criteria not described	0.02 - 0.09	Fowler et al. (1971)
Wall and plenum insulation	5		0.01 - 0.47	
Pipe insulation	6		0.02 - 0.09	
Housing insulation	1		0.20	
Acoustic ceiling installer	12	phase contrast optical microscopy; length: diameter > 3:1; length > 5 µm; diameter < 3 µm; transmission electron microscopy	0.0028 (0 - 0.0006)	Esmen et al. (1982)
Duct installation				
- pipe covering	31		0.06 (0.0074 - 0.38)	
- blanket insulation	8		0.05 (0.025 - 0.14)	
- wrap around	11		0.06 (0.030 - 0.15)	
Attic insulation (fibrous glass)				
- roofer	6		0.31 (0.073 - 0.93)	
- blower	16		1.8 (0.67 - 4.8)	
- feeder	18		0.70 (0.06 - 1.5)	
Attic insulation (mineral wool)				
- roofer	9		0.53 (0.041 - 2.03)	
- blower	23		4.2 (0.50 - 15)	
- feeder	9		1.4 (0.26 - 4.4)	
Installer of building insulation	31		0.13 (0.013 - 0.41)	

the installation of fibrous glass insulation in attics have ranged up to 1.8 fibres/cm^3; mean concentrations during the application of mineral wool insulation under similar circumstances have been as high as 8.2 fibres/cm^3. During the installation of rock wool blankets in very confined spaces on board ships, concentrations have been less than 0.65 fibres/cm^3 (Marconi et al., in press). It should be noted that the time-weighted average exposure of insulation workers is probably considerably less than these mean concentrations during application, as insulators work with MMMF materials from < 10 to 100% of their time, depending on their employment and the type of construction site (Fowler et al., 1971). The majority of joiners and carpenters may use from 0.5 to 15% of their working hours on MMMF work (Schneider, 1984). On the basis of estimates made by Esmen et al. (1982), time-weighted average exposures may exceed 1 fibres/cm^3 only for workers insulating attics with mineral wool (TWA = 2 fibres/cm^3; range, 0.29 - 6.3 fibres/cm^3). Time-weighted average exposures for workers using fibrous glass in attics or employed in building insulation were estimated to be 0.7 fibres/cm^3 (range, 0.42 - 1.2 fibres/cm^3) and 0.11 fibres/cm^3 (range, 0.08 - 0.2 fibres/cm^3), respectively.

Air at construction sites and in certain other industrial and domestic environments may contain substantial amounts of non-fibrous dust. The gravimetric concentration of fibres in total dust samples has been determined by optical microscopy for a range of user industries (Schneider, 1979). The ratio weight of all fibres of all sizes in total dust samples/weight of total dust had a geometric mean of 0.14, with 90% between 0.06 and 0.33.

Howie et al. (1986) recently investigated fibre release from filtering facepiece respirators containing "superfine" MMMF. These respirators are often used for the protection of workers in dusty occupational environments. Following the use of several different types of these respirators for either 15 or 40 min, downstream respirable fibre concentrations ranged from not detectable (< 0.1 fibres/cm^3) to 200 fibres/cm^3, using the WHO reference PCOM method of monitoring.

5.1.1.2 Ambient air

Few data are available concerning concentrations of MMMF present in the general environment. The fibrous glass content of several samples of outdoor air was determined in a study by Balzer et al. (1971), which was designed principally to investigate the possible erosion of fibres from air transmission systems lined by fibrous glass. Mean concentrations on the rooftops of 3 buildings on the University of California's Berkeley Campus, determined by PCOM and petrographic microscopy, were about 2.7×10^{-4} fibres/cm^3 (range, $< 5 \times 10^{-5}$ - 1.2×10^{-3}

fibres/cm^3). Levels at other sites averaged 4.5×10^{-3} fibres/cm^3, the lowest levels being 4×10^{-4} fibres/cm^3. Fibres other than MMMF may have been included in the results of these analyses.

Balzer (1976) reported the results of a further study in which MMMF levels were determined by light and electron microscopy in 36 samples of ambient air from various locations in California (Berkeley, San Jose, Sacramento, the Sierra Mountains, and Los Angeles). Fibre counts were determined by combining the light microscopic count of fibres greater than 2.5 μm in diameter and the electron microscopic count of fibres less than, or equal to, 2.5 μm in diameter. Levels of glass fibres averaged 2.6×10^{-3} fibres/cm^3 and accounted for approximately 1/3 of the total fibrous material in the samples. However, the significance of these results is uncertain, since the methods of sampling and analysis and the results were not well described in the published account of this study.

Mean airborne glass fibre concentrations measured by analytical TEM ranged from 4×10^{-5} fibres/cm^3 at 1 rural location to 1.7×10^{-3} fibres/cm^3 in 1 out of 3 cities in the Federal Republic of Germany (1981-82; 9 - 21 samples at each location). These levels were 3 - 40% of the asbestos concentrations; median diameters of the glass fibres ranged from 0.25 μm to 0.89 μm and median lengths from 2.54 to 3.64 μm (Höhr, 1985).

5.1.1.3 Indoor air

On the basis of analysis using PCOM (Balzer et al., 1971; Esmen et al., 1980) and calculation of the "glass" content of collected particulate (Cholak & Schafer, 1971), it has generally been concluded that the contribution of fibrous-glass-lined air transmission systems to the fibre content of indoor air is insignificant. Using very conservative assumptions, Esmen et al. (1980) estimated that the level of fibres in occupied spaces is of the order of 0.001 fibres/cm^3 during the first day of operation of air transmission systems with medium grade fibrous glass filters and essentially the same as ambient levels during the remaining filter life.

Generally, concentrations of MMMF in indoor air are 100 - 1000 times less than those in the occupational environment. Several studies in which airborne fibre levels in indoor air were measured by PCOM with polarization equipment have been carried out in Denmark (Table 12). Although fibres other than MMMF may have been included in these analyses, birefringent fibres (mostly organic) were counted separately. Typical levels, as well as total dust concentrations measured in one of the studies (Nielsen, 1987), are shown in Table 13.

Table 12. Respirable MMMF fibre concentrations in the indoor environment (static samples)

Site	Number of buildings	Respirable fibre levels (x 10^{-6} fibres/cm^3)[a] Arithmetic mean and range of individual samples	Comments	Reference
Random sample of mechanically ventilated schools	11	60 (0 - 240)		Schneider (1986)
Random sample of kindergardens (sites with reported indoor air problems were excluded)	5	110 (60 - 160)[b]	MMMF ceiling boards with water-soluble binder	Rindel et al. (1987)
	3	100 (43 - 150)[b]	Resin binder	
	4	40 (10 - 70)[b]	No MMMF ceiling boards	
140 rooms (day-care centres, schools, offices) selected at random, but with MMMF ceiling boards		260 (0 - 1330)	Water-soluble binder, untreated surface	Nielsen (1987)
		75 (0 - 425)	Water-soluble binder, untreated surface, varnished	
		70 (0 - 180)	Water-soluble binder, untreated surface	
		30 (0 - 250)	Resin binder	
		40 (0 - 85)	Resin binder	
		250 (0 - 1070)	Resin binder	
		140 (0 - 820)	Resin binder	
		25 (0 - 80)	Resin binder	
		40 (0 - 480)	No MMMF	
Sites with reported indoor climate problems	6	range, 230 - 2900	One observation (0.084 fibres/m^3)	Schneider (1986)

[a] Phase contrast optical microscopy with polarization. Length $\geq$ 5 μm; diameter $\leq$ 3 μm; aspect ratio $\geq$ 3:1.

Table 13. Average concentrations of total dust and fibres other than MMMF in indoor environments (static samples)[a]

Site	Number of measurements	Total dust (mg/m^3)	Organic and other birefringent fibres ($fibres/cm^3$)
Day-care centres	49	0.26	0.25
Schools	11	0.16	0.064
Offices	39	0.13	0.025

[a] From: Nielsen (1987).

5.1.2 Water supplies

Data on the MMMF content of water supplies are not available. However, glass fibres have been identified by optical microscopy in samples of sewage sludge from 5 cities in the USA (Bishop et al., 1985).

5.2 Historical Exposure Levels

In an analysis of the data from the large European epidemiological study referred to previously, the production history of each factory was classified into the following 2 distinct technological phases: (a) an early phase during which a batch production system was in use and/or no oil added during production; and (b) a late phase during which modern production techniques were used and oil was added. A third phase, intermediate between the early and late phases, was also identified at some plants where a mixture of production types or techniques operated.

The airborne fibre levels in the early technological phase in the slag wool/rock wool industry were probably substantially higher than in the late phase, the corresponding levels in the intermediate phase being between these two levels (Cherrie & Dodgson, 1986). However, the pattern in the glass wool plants was somewhat different.

The basis of these differences rests primarily with changes over time in the nominal fibre diameters of the MMMF being produced, and the addition of oil. In the early period in the slag wool/rock wool industry, the lack of oil and fibres of relatively fine nominal diameter (3 - 6 μm) contributed to higher airborne fibre levels. On the other hand, in the glass wool production industry, it is likely that there was little change in airborne fibre levels between the early and late phases. In the early phase, oil was not added during production, and the nominal diameter of the fibres produced was relatively large (10 - 25 μm). In the late phase, oil was added during production, and the nominal diameter of the fibres produced was smaller (5 - 7 μm). The effects of changes in other factors including ventilation and production rates were judged to be less.

Combined experimental and modelling procedures have been used to estimate past exposure levels in the European MMMF plants (Dodgson et al., in press). The model is still being developed and should be considered to be preliminary (Cherrie et al., in press). It indicates that mean airborne fibre concentrations for rock wool plants could have been about 1 - 2 fibres/cm^3, during the early technological phase, with levels of about 10 fibres/cm^3 for the dustiest work. Current exposure levels are at least one order of magnitude lower. Corresponding estimates indicate that the mean airborne fibre levels for the glass wool plants during the early technological phase were little different from current concentrations (about 0.1 fibres/cm^3 or less).

The validity of the model has been investigated in an experimental simulation of rock wool production in the early phase (Cherrie et al., in press). The effects of added oil on the airborne fibre levels during experimental production were assessed together with the effects of workers handling batches of material. Addition of oil resulted in a 3- to 9-fold reduction in the airborne fibre levels in different situations. The time-weighted average concentrations were 1.5 fibres/cm^3 with oil added and about 5 fibres/cm^3 without oil. These values agreed reasonably well with the predictions from the model. There was no substantial difference in fibre concentrations between batch and continuous operation. However, it is possible that the batch handling was not well simulated.

5.3 Exposure to Other Substances

Polyaromatic hydrocarbons (PAHs) and other combustion products were present in the working environment, particularly in the early technological phase of the slag wool/rock wool industry in Europe, where cupolas were used extensively. In one plant, the use of an olivin, potentially contaminated with a natural mineral fibre with a composition similar to tremolite was reported (Cherry & Dodgson, 1986). The probable average exposure in the pre-production area was estimated to be 0.1 fibres/cm^3. The European slag wool plants used copper slags, possibly liberating arsenic compounds into the working environment.

In MMMF plants in the USA, other airborne contaminants have been measured in areas not necessarily representative of occupational exposure or in breathing zones (IARC, in press). The results, obtained periodically over the years 1962-87, are shown in Table 14. It was emphasized that these measurements only verify the presence of other contaminants in the plants and cannot be used to estimate the degree of exposure.

Table 14. Exposure to other airborne contaminants in MMMF plants in the USA[a]

Contaminant	Range of concentrations
Asbestos	0.02 - 7.5 fibres/cm^3
Arsenic	0.02 - 0.48 $\mu g/m^3$
Chromium (insoluble)	0.0006 - 0.036 mg/m^3
Benzene-soluble organics	0.012 - 0.052 mg/m^3
Formaldehyde	0.06 - 20.4 mg/m^3
Silica (respirable)	0.004 - 0.71 mg/m^3
Cristobalite (respirable)	0.1 - 0.25 mg/m^3

[a] From: IARC (in press).

6. DEPOSITION, CLEARANCE, RETENTION, DURABILITY[a], AND TRANSLOCATION

6.1 Studies on Experimental Animals

Because of the tendency of fibres to align parallel to the direction of airflow, the deposition of fibrous particles in the respiratory tract is largely a function of fibre diameter, length and aspect ratio being of secondary importance.

Since most of the data on deposition of various MMMF have been obtained in studies on rodents, it is important to consider differences between rats and man in this regard. Comparative differences between rat and man can best be evaluated using the aerodynamic equivalent of fibres. The ratio of the absolute diameter to aerodynamic diameter is approximately 1:3. Thus, a fibre measured microscopically to have a diameter of 1 μm would have a corresponding aerodynamic diameter of approximately 3 μm. In addition, any curvature of the fibres may have the effect of increasing the effective aerodynamic diameter.

A comparative review of the regional deposition of particles in man and rodents (rats and hamsters) has been presented by US EPA (1980). The relative distribution between the tracheo-bronchial, and pulmonary regions of the lung in rodents followed a pattern similar to human regional deposition during nose breathing for insoluble particles of less than 3 μm mass median aerodynamic diameter (approximately 1 μm diameter or less). Fig. 4 and 5 illustrate these comparative differences. As can be seen, particularly for pulmonary deposition, the percent deposition of particles is considerably less in the rodent than in man. These data indicate that, while particles of 5 μm aerodynamic diameter or greater may have significant deposition efficiencies in man, the same particles will have extremely small deposition efficiencies in the rodent. More recent work by Snipes et al. (1984) indicates that, with particles above 3 μm aerodynamic diameter, the probability of reaching the pulmonary region falls off rapidly, and that particles of 9 μm aerodynamic diameter have an extremely small probability of reaching the pulmonary region in rats and guinea-pigs.

Because of their length, fibres are more likely than spherical particles to be deposited by interception, mainly at bifurcations. Available data also indicate that pulmonary penetration of curly chrysotile fibres is less than that for straight solid amphibole fibres. However, in studies on rats, in which the deposition of curly versus straight glass fibres was

---

[a] Durability of MMMF in tissues is a function of their physical and chemical characteristics.

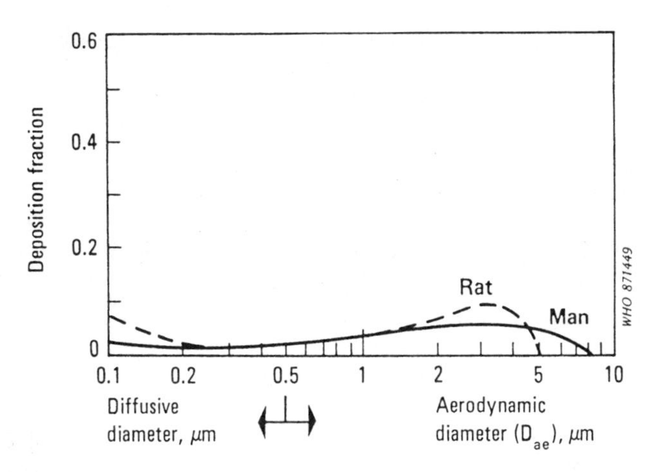

Fig. 4. Tracheobronchial deposition of inhaled monodisperse aerosols in man and rodent. From: US EPA (1980).

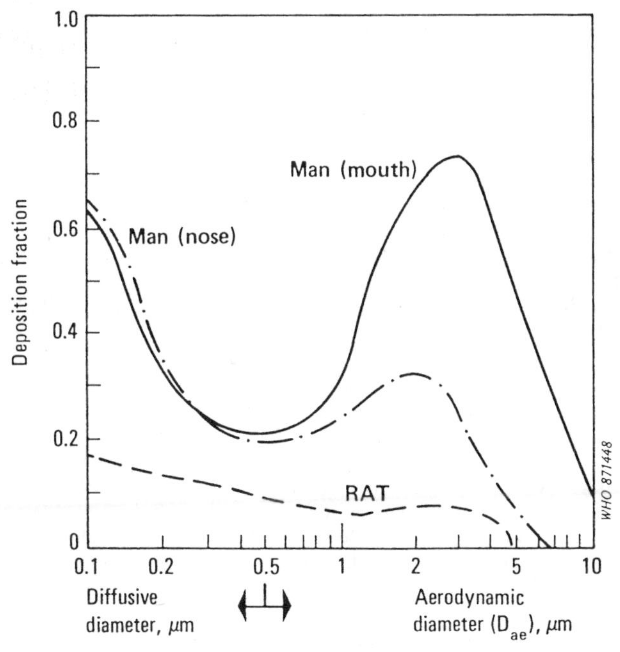

Fig. 5. Pulmonary deposition of inhaled monodisperse aerosols in man and rodent. From: US EPA (1980).

examined, the difference was less apparent (Timbrell, 1976). Clearance from the peripheral respiratory tract is principally a function of fibre length, with fibres shorter than 15 - 20 μm in length being cleared more efficiently than longer ones.

The results of available studies concerning the deposition, clearance, retention, durability, and translocation of MMMF in animal species are presented in Table 15. As for all fibrous particles, deposition of MMMF in the respiratory tract is determined principally by fibre diameter. In studies conducted by Morgan et al. (1980), alveolar deposition of glass fibres in rats was much less than that for asbestiform minerals, because of the larger aerodynamic diameter of MMMF. Deposition of asbestiform and glass fibres, the aerodynamic diameters of which ranging from 1 to 6 μm), was greatest for fibres with an aerodynamic diameter of 2 μm (Morgan et al., 1980). However, there was some deposition of glass fibres with aerodynamic diameters of between 3 and 6 μm. For fibres of constant diameter, alveolar deposition decreased with increasing fibre length. Morgan & Holmes (1984b), comparing their results in rats with those concerning fibre deposition in other species, suggested that glass fibres with diameters greater than 3.5 μm (aerodynamic diameter ~10 μm) are unlikely to be respirable in man by nose breathing (Timbrell, 1965; Morgan et al., 1980; Hammad et al., 1982; Hammad, 1984). Fibre deposition in various lobes of the lungs of rats varied over a narrow range (5.34 - 8.38%) following inhalation of ceramic fibres. Fibre size distributions in the various lobes, based on analysis by phase contrast optical microscopy, were not significantly different (Rowhani & Hammad, 1984). These results are similar to those observed for glass wool, rock wool, and glass microfibre by Le Bouffant et al. (in press).

According to Griffis et al. (1981), there is a fairly rapid decline in the lung content of glass fibres, immediately following deposition, presumably because of mucociliary clearance ($t\frac{1}{2}$ in rats ~1 day). This is followed by a much slower phase assumed to represent alveolar clearance ($t\frac{1}{2}$ in rats ~44.3 days) (Friedberg & Ullmer, 1984). Short fibres are efficiently cleared by alveolar macrophages (Morgan et al., 1982a; Morgan & Holmes, 1984b; Bernstein et al., 1984); in rats, more than 80% of glass fibres of less than 5 μm in length were cleared by one year ($t\frac{1}{2}$ ~60 days). However, macrophage-mediated clearance appeared to be ineffective for fibres with lengths of 30 and 60 μm (Morgan & Holmes, 1984b).

The degradation of MMMF in lung tissue consists of reduction of diameter with time, together with pitting and erosion of the surface visible on TEM (Morgan & Holmes, 1986). Observations for up to 2 years after intratracheal instillation in rats showed a wide range in the durability of various MMMF (Bellmann et al., in press). Morgan & Holmes (1986) noted that relatively

Table 15. Deposition, durability, clearance, retention, and translocation of MMMF in experimental animals

Species	Study protocol[a]	Results	Reference
Alderley Park (strain 1) SPF rat	Inhalation for 2 - 3 h of samples of glass fibres and asbestiform materials at concentrations of 0.40 - 1.63 g/litre; aerodynamic diameter of fibres ranged from 1 to 6 µm; animals killed immediately or 2 days following exposure	Decrease in alveolar deposition with increasing fibre length; maximum deposition for fibres with aerodynamic diameter of 2 µm; small but significant deposition of fibres with aerodynamic diameters of 3 - 6 µm	Morgan et al. (1980); Morgan & Holmes (1984b)
Fischer 344 SPF male rat	Inhalation of 31 - 52 mg glass fibres/m^3 (2/3 respirable; CMD, 0.11 µm; CML, 8.3 µm), 6 h/day, for 1, 2, 4, or 5 days	41 - 48% of lung burden cleared between daily exposures (i.e., half-time for early clearance, 1 day)	Griffis et al. (1981)
Beagle dog	Inhalation for 1 h of unspecified concentration of glass fibres (CMD, 0.15 µm; CML, 5.4 µm); animals killed 4 days after exposure	77% of body burden excreted in 4 days, mainly (> 96%) in faeces (half-time 2 days); 5 - 17% deposited in deep lung; respiratory tract deposition, 45 - 64%	Griffis et al. (1983)
Male albino rat	Inhalation of ~300 fibres/cc (CMD, 1.1 - 1.3 µm; CML, 8 - 20 µm); animals killed 5 days after exposure	Increased alveolar retention of fibres with decreasing diameter and decreasing length (maximum 7.6% for fibres < 0.5 µm in diameter and 21 µm in length)	Hammad et al. (1982)

Table 15 (contd).

Species	Study protocol[a]	Results	Reference
Syrian golden hamster	Intratracheal instillation of 0.2 mg sized glass fibres (3 µm in diameter; length, 10 - 100 µm); animals killed serially up to 8 months after administration	Preferential clearance of shorter fibres; ferruginous bodies around fibres > 10 µm in length only; time of onset of body formation decreased with increasing fibre length; proportion of coated fibres increased over a period of about 3 - 4 months and then declined after 5 months; coating did not prevent dissolution of fibres; longer fibres dissolved more rapidly than shorter ones; fibres dissolved more rapidly in trachea than in rat lung	Holmes et al. (1983); Morgan & Holmes (1984b)
Alderley Park (strain 1) SPF rat and Syrian golden hamster	Intratracheal instillation in rats (0.5 mg) or hamsters (0.4 mg) of rock wool (CMD, 1.1 µm; CML, 28 µm); animals killed serially up to 18 months after administration	Rock wool fibres dissolved more slowly in the lung than glass fibres; diameter essentially unchanged after 18 months; in hamsters, coating of fibres < 2 µm in diameter after 2 months	Morgan & Holmes (1984a,b)
Alderley Park (strain 1) SPF rat	Intratracheal instillation of 0.5 ml of 6 samples of glass fibres (1.5 or 3 µm in diameter; length, 5 - 60 µm); animals killed serially up to 18 months after administration	Fibres with diameter < 3 µm and lengths < 10 µm efficiently cleared (presumably by macrophage-mediated processes); fibres 30 and 60 µm in length not cleared to a significant extent over 1 year; longer fibres dissolved more rapidly in a non-uniform manner leading to fragmentation; shorter fibres dissolved more slowly and uniformly	Morgan et al. (1982); Morgan & Holmes (1984b)

Table 15 (contd)

Guinea-pig	Intrapleural injection of 25 mg commercial glass fibre (diameter, 0.05 - 0.99 µm); animals killed 6 weeks after administration	Fine structure of coated glass fibres in lung identical to that of asbestos bodies	Davis et al. (1970)
Sprague Dawley rat	Inhalation for 10 or 30 h of "TEL" glass fibres at 86 mg/m^3; animals killed 16 - 18 h, 7, 14, 30(33), 60, or 90 days after exposure	Mean half-time for elimination 44.3 days (assuming first order)	Friedberg & Ullmer (1984)
Fischer 344 SPF rat	Inhalation of 10 mg/m^3 rock wool, glass wool (with and without resin), and glass microfibre for 7 h/day, 5 days/week, for 1 year; animals killed immediately or at unspecified periods after exposure	Glass microfibres more susceptible to etching in lung tissue than glass or rock wool	Johnson et al. (1984a)
Male rat	Inhalation of mineral wool (CMD, 1.2 um; CML, 13 µm) or ceramic fibres (CMD, 0.7 µm; CML, 9 µm) (~300 fibres/cc), 6 h/day, for 5 or 6 days; animals killed 5, 30, 90, 180, or 270 days after exposure	Ceramic fibres more persistent than mineral wool in lung tissue	Hammad (1984)
Wistar IOPS AF/Han rat	Inhalation for 1 or 2 years of resin-free glass wool or rock wool (Saint Gobain), or microfibres (JM 100) (~5 mg/m^3); animals killed immediately or serially for periods up to 16 months after exposure	Fibre concentrations in the air and lung much greater for microfibre (JM 100) than for glass and rock wool; retention and translocation less for rock and glass wool fibres than for microfibres; lengths of glass and rock wool fibres retained in the lungs shorter than those of airborne fibres, whereas glass microfibre sizes in lung and air similar; generally, fibres in the lymph nodes shorter than those in the lungs	Le Bouffant et al. (1984, in press)

Table 15 (contd).

Species	Study protocol[a]	Results	Reference
Male Fischer 344 SPF rat	Intratracheal instillation of 2 or 20 mg glass fibres (1.5 µm in diameter; length, 5 or 60 µm); animals killed serially up to 2 years after administration; intratracheal instillation of 0.5 mg glass fibres (1.5 µm in diameter; length, 5 or 60 µm) or tracheal inhalation once per week, for 10 weeks	Short fibres cleared efficiently by macrophages with fewer than 10% remaining after 500 days; longer fibres dissolved more rapidly (over 50% in 18 months), possibly due to partial phagocytosis	Bernstein et al. (1980, 1984)
Pathogen-free adult albino male rat	Inhalation of ceramic fibre (CMD, 0.53 µm; CML, 3.7 µm) (~709 fibres/cm^3), 6 h/day, for 5 days; animals killed 5 days after exposure; analysis by PCOM[b]	Deposition of fibres varied over a narrow range (5.43% for the right diaphragmatic lobe to 8.38% for the right apical lobe); fibre burden for all lobes weight dependent; no significant difference in fibre size distributions in various lobes	Rowhani & Hammad (1984)

Table 15 (contd).

Female Wistar rat	Intratracheal instillation of 2 mg microfibre (JM 104-E glass: 90% of fibre diameters < 0.28 µm, 90% of fibre lengths < 5.8 µm; JM 104/475: 90% of fibre diameters < 0.4 µm, 90% of fibre lengths < 8.4 µm; acid-treated microfibre (JM 104-E glass: 90% of fibre diameters < 0.46 µm, 90% of fibre lengths < 6.3 µm); rock wool (90% of fibre diameters < 5.6 µm, 90% of fibre lengths < 67 µm); ceramic wool (90% of fibre diameters < 4.2 µm, 90% of fibre lengths < 177 µm) with animals killed serially up to 2 years after administration	Half lives for fibres > 5 µm in length: JM 104/475, 3500 days (similar to crocidolite; change in size not detected); JM 104-E glass, 55 days (increase in mean length and diameter in first 6 months); acid-treated JM 104-1974, 14 days; rock wool, 283 days; ceramic wool, 780 days	Bellmann et al. (in press)

a CMD = Count median diameter.
CML = Count median length.

b PCOM = Phase contrast optical microscopy.

thick MMMF become thinner on dissolution and may assume dimensions that resemble those of the fibrous amphiboles. However, the *in vivo* studies in this report, together with the *in vitro* observations of Klingholz & Steinkopf (1984), suggest that the leaching of alkaline ions that accompanies fibre dissolution results in rapid changes in the nature of the surface of the fibre, and loss of substance, so that the cell/fibre interface may be altered. In studies on rats, long glass fibres degraded more rapidly than short ones (Morgan et al., 1982; Holmes et al., 1983; Bernstein et al. 1984; Hammad, 1984). It has been postulated that this may be because the small (short) fibres are completely engulfed by macrophages and therefore "protected" from the extracellular environment, which contains more fluids and enzymes that might facilitate degradation.

Although ferruginous bodies are not observed in rats exposed to asbestos or glass fibres, in an early study on guinea-pigs, it was demonstrated that glass fibres could be coated in the lung to form bodies with a fine structure identical to that of asbestos bodies (Davis et al., 1970). In hamsters, various proportions of the fibres in lung tissue greater than 10 μm in length became coated to form ferruginous bodies; the time of onset of formation decreased with increasing fibre length (Holmes et al., 1983). In studies involving exposure for 90 days (0.4 mg glass fibres/litre), by Lee et al. (1981), ferruginous bodies were first detected in hamsters by the 6th month following exposure and in guinea-pigs by the 12th month. The coating of ferruginous bodies appears to be discontinuous and does not prevent the leaching and fragmentation of long glass fibres (Morgan & Holmes, 1984b).

Available data also indicate that rock wool may be more persistent in lung tissue than fibrous glass (Johnson et al., 1984a; Morgan & Holmes, 1984a,b). Following intratracheal instillation in rats, the diameter of rock wool fibres in the lung was essentially unchanged after 18 months; however, the ends of the fibres were perceptibly thinner. The results of an additional study on rats indicated that ceramic fibres were more persistent in lung tissue than rock wool, even though the mean diameter of the mineral wool fibres was greater than that of the ceramic fibres. After 270 days, about 25% of ceramic fibres were still present in lung tissue compared with 6% of mineral wool fibres (Hammad, 1984).

Lee et al. (1981) reported that dust particles were present in the tracheobronchial lymph nodes after 50 days exposure to 0.4 mg glass fibre/litre. Following inhalation of glass wool, rock wool, or glass microfibre at 5 mg/m^3 for 1 year, fibre concentrations in the tracheobronchial glands of rats were low (range of means in males and females, 0.1 - 0.8 g/kg dry tissue), except for microfibres (JM 100), for which mean levels were 4.8 and 3 g/kg dry tissue, in males and females, respectively (Le Bouffant et al., 1984) (For comparison, mean concentrations of glass and rock wool in lung tissue ranged from 0.4 to 2.8 g/kg dry weight; mean levels of microfibre were 5.8 - 12.7 g/kg dry lung tissue). Levels of the various fibre types in the diaphragm were essentially zero (Le Bouffant et al., 1984). Generally, levels of fibres in the lymph nodes of animals exposed for 2 years were higher than those after 1 year of exposure (Le Bouffant et al., in press). Bernstein et al. (1984) reported that, following intratracheal instillation of size-selected glass fibres of diameters of 1.5 μm and lengths of either 5 μm (short) or 60 μm (long), short fibres were quickly cleared to the regional lymph nodes, whereas the long fibres were not. Even 18 months after exposure, there were no significant quantities of fibres in the lymph nodes of the groups exposed to long fibres, whereas there were many still present in the animals exposed to short fibres. Spurny et al. (1983) reported the presence of glass fibres in the spleen of rats, 2 years after intratracheal implantation. However, few experimental details were provided. Recovery of fibres from all organs examined 90 days after intrapleural injection of fibrous glass (20 mg JM 104) was reported by Monchaux et al. (1982).

6.2 Solubility[a] Studies

The results of studies on the solubility of MMMF in physiological fluids *in vitro* are presented in Table 16. Since it is impossible to reproduce *in vitro* the various conditions of pH and concentrations of complexing agents in the intra- and extracellular environment of the lung, the results of dissolution studies should be interpreted with caution. However, on the basis of these investigations, it appears that dissolution is a function not only of the fibre characteristics, such as chemical composition and size, but also of the nature of the leaching solution. In one of these investigations, MMMF were more soluble than the amphiboles and chrysotile (Spurny, 1983). However, there is a wide range in the solubility of the various MMMF, e.g., 30- to 40-fold in the study of Leineweber (1984). For example, Leineweber (1984) reported that the stability of various MMMF in physiological saline was as follows: glass fibre (1 type) > refractory fibre > mineral wool > glass fibre (3 types). However, in distilled water, the order was significantly different: refractory fibre > glass fibre (2 types) > mineral wool > glass fibre (2 types).

---

[a] Solubility of MMMF is their behaviour in various fluids. In general, the term is restricted to observations *in vitro*, since the persistence of fibres *in vivo* is a function of several factors including solubility.

Table 16. The solubility of MMMF

Study protocol	Results	Reference
Analysis by scanning electron microscopy, microprobe analysis, X-ray diffraction, X-ray fluorescence spectroscopy and mass spectroscopy of UICC chrysotile or crocidolite glass fibres and mineral wool in 5 solutions (2N HCl, 2N H_2SO_4, 2N NaOH, water, and blood-serum)	Surface porosity was altered and alkali elements were increased; in general, order of stability: amphiboles > glass fibres > other MMMF > chrysotile; in some cases, succession amphiboles > chrysotile > MMMF; fine diameter fibres degrade more rapidly than coarse fibres	Spurny (1983); Spurny et al. (1983)
Analysis by atomic absorption spectrometry, scanning electron microscopy and energy dispersive X-ray analysis of asbestos, rock wool, slag wool, basalt wool, and glass wool (without binders) in deionized distilled water, KOH (48%), H_2SO_4 (48%), Ringer's solution, and Gamble's solution	MMMF softened in physiological solutions and gel forms that is removed in layers; rate of dissolution in Gamble's solution: 10 ng/cm^2 per h (i.e., fibres with average diameter of 3 µm completely dissolved in < 5 years; those with diameter < 1 µm dissolved in < 1 year); asbestos stable in physiological solutions	Forster (1984)
Analysis by atomic absorption spectrometry, electron probe, scanning electron microscopy, and energy dispersive X-ray analysis of mineral wool, glass wool, rock wool, and basalt wool (without binders) in Gamble's solution	Gel formation on the surfaces of all MMMF within a few weeks, particularly for those with higher alkaline content; beneath the gel layer, areas of local corrosion ("pitting") lead to loss of structural integrity and breakage into shorter lengths	Klingholz & Steinkopf (1984)

Table 16 (contd).

Analysis by BET gas absorption, scanning electron microscopy, and energy dispersive X-ray spectrometry of 4 samples of fibrous glass, mineral wool, and refractory fibre in physiological saline and distilled water	40-fold range of durability in saline; order of durability: 1 sample of glass fibre > refractory fibre > mineral wool > 3 samples of glass fibre; order in distilled water varied; correlation between the solubility and total weight percentage alkalis; surface deposits observed	Leineweber (1984)
Determination of mass and sodium content by flame atomic absorption spectroscopy of glass fibres in tris buffered saline or serum stimulant	Greater mass dissolved in serum stimulant than in buffered saline; some sodium preferentially dissolved in both solvents	Griffis et al. (1981)
Analysis (inductive coupled plasma method) for silicon, boron, and potassium; weight determination and examination by scanning electron microscopy	MMMF much more soluble (dissolution time for glass or slag wool $\leq$ 2 years, for E glass wool (JM 104) and refractory fibre, 5 - 6.5 years) than natural mineral fibres (dissolution time > 100 years)	Scholze & Conradt (in press)

On the basis of available data, it appears that there is a correlation between solubility in physiological solution and the alkaline content of MMMF (Klingholz & Steinkopf, 1984; Leineweber, 1984) and that fibres of fine diameter degrade more rapidly than coarse ones (Spurny et al., 1983; Forster, 1984).

7. EFFECTS ON EXPERIMENTAL ANIMALS AND *IN VITRO* TEST SYSTEMS

There are several factors that should be considered when evaluating experimental data on the biological effects of MMMF. Most importantly, MMMF should not be considered as a single entity, except in a very general way. There are substantial differences in the physical and chemical properties (e.g., fibre lengths and chemical composition) of the fibres, even within each of the broad classes of MMMF (continuous filaments, insulation wools, refractory fibres, and special purpose fibres), and it is expected that these would be reflected in their biological responses. Finally, the fibres used for specific research protocols may be altered to fit the needs of the study, and the results may not necessarily represent the true biological potential of the parent material. Therefore, the potential "toxicity" of each fibrous material must be evaluated individually.

Notwithstanding the above comments, there are certain characteristics of MMMF (and other fibres) that are important determinants of effects on biological systems. The most important of these appear to be fibre size (length, diameter, aspect ratio), *in vivo* persistence[a] and durability[a], chemical composition, surface chemistry, and number or mass of fibres (dose). In addition, it is imperative to understand that each of the above characteristics cannot be viewed in isolation from the others. For example, it is clear that two types of fibres of identical dimensions could react differently within the host if they differ in one of the other characteristics. Some of the apparently inconsistent experimental results of studies reported in this section may be a function of these factors.

---

[a] The term persistence refers to the ability of a fibre to stay in the biological environment, where it was introduced. The term is of particular use in inhalation and intratracheal instillation studies, because a large percentage of the fibres that reach the lung are removed by the mucociliary apparatus and relatively few are retained (persist). The length of time that MMMF persist in the tissue is also a function of their durability, which is directly related to their chemical composition and physical characteristics. The term solubility, as used here, relates to the behaviour of MMMF in various fluids. In general, the term solubility is more appropriate for use in *in vitro* than in *in vivo* studies, because, in tissue, the degradation of fibres is a function of phenomena more than just their solubility.

Other considerations relate to the extrapolation of experimental findings for hazard assessment in man. It appears that if a given fibre comes into contact with a given tissue in animals producing a response, a similar biological response (qualitative) might be expected in human beings under the same conditions of exposure. There is no evidence that the biological reaction of animals and human beings differ. However, there may be quantitative differences between species, some being more "sensitive" than others.

There has been a great deal of debate concerning the "relevance" of various routes of exposure in experimental animal studies to hazard assessment in man. The advantages and disadvantages of each of these routes are discussed in the following sections. It was the consensus of the Task Group that administration by each route has its place in the study of fibre toxicity, but that it is inappropriate to draw conclusions on the basis of the results by one route of administration only. This does not mean that all routes are required for the study of the toxicity of a given fibre.

Each route cannot be discussed in detail here, but some general observations can be made. First, positive results of an inhalation study on animals have important implications for hazard assessment in man. Strong scientifically based arguments would need to be made to dissuade one from the relevance of such a finding to man. Conversely, a negative inhalation study does not necessarily mean that the material is not hazardous for human beings, unless it is clear that the number of fibres that actually reached the target tissue (in this case the lung) was comparable to the positive control. Rats, being obligate nose breathers, have a greater filtering capacity than human beings. However, if it were demonstrated that the "target tissue" was adequately exposed and that a biologically important response was not noted, then such a result would be of value for hazard assessment in human beings.

In contrast to inhalation studies, a negative result in a properly conducted intratracheal study would suggest that a given type of fibre may not be hazardous for parenchymal lung tissue. A positive result, in such a case, would require further study before assessing the hazard for man, since the normal filtering capacity of the respiratory tract has been bypassed, and there is often a non-random distribution in the lung, i.e., "bolus" effect. In other words, in intratracheal instillation studies, the lung can be exposed to material with which it normally would never come into contact. However, pulmonary clearance mechanisms are still intact, though their efficiency may be compromised.

The results of studies involving intrapleural instillation (injection or implantation) and intraperitoneal injection should be viewed in a similar way to those of investigations involving intratracheal instillation studies. With these methods, both filtering and clearance mechanisms are confused. However, such studies may be more "sensitive" than inhalation studies, because a higher number of fibres (> 1.0 μm in diameter) can be introduced than can be in conventional inhalation studies. Therefore, a negative result in such a case would be highly relevant in terms of hazard assessment. In contrast, a positive result should be confirmed by further investigation involving other routes of exposure before indicting the particular type of fibre as a human hazard in terms of the tissue involved (mesothelioma). It should be noted that both of these routes are primarily relevant to the mesothelium and do not necessarily predict what happens in the pulmonary parenchyma or airways.

In vitro short-term studies, e.g., cytotoxicity, cytogenicity, and cell transformation studies, have been, and are, important in understanding the mechanisms of action of fibres. The results of such studies are of value in the overall assessment of fibre toxicity but should not be used alone for hazard assessment.

Solubility[a] studies play an important part in the understanding of the behaviour of fibres behave in various media. However, several of the mechanisms that determine the persistence of fibres in biological tissues are not taken into account. Therefore, while an important adjunct in the overall study of fibres, they are of less importance in hazard identification.

While it is not the purpose of this publication to outline a specific set of protocols for the study of the biological effects of newly developed MMMF, it may provide some guidance concerning those aspects that should be addressed in experimental studies to investigate the potential toxicity of fibres.

7.1 Experimental Animals

7.1.1 Inhalation

Exposure conditions in inhalation studies approach most closely the circumstances of human exposure and are most relevant for the assessment of risks for man. Information on the design and results of studies in which animals were exposed to MMMF by inhalation is summarized in Table 17. Such investigations have been conducted on several species including the rat, mouse, hamster, guinea-pig, and baboon.

---

[a] See footnote [a] on p. 65.

Table 17. Inhalation studies

Species	Number	Protocol	Results and comments	Reference
Male A-strain mouse	12 animals each in 3 treatment groups; 12 vehicle- and 12 unexposed control animals	300 mg crushed "fibreglass insulation" (80% of fibres were 2.5 µm wide and 6 - 11 um long), in bedding, every 3 days, for 30 days; 3 days later, ip administration of either retinyl palmitate (50 mg/kg body weight), ascorbic acid (400 mg/kg body weight), or both; animals sacrificed 90 days after first exposure to fibrous glass	Neoplastic, hyperplastic, and metaplastic lesions in the bronchi and near the bronchiole-alveolar junctions in the lung; administration of either or both vitamins resulted in significantly fewer neoplastic and hyperplastic lesions; slightly higher number of metaplastic nodules in group exposed to retinyl palmitate; control animals had 5 times as many particles in the lung as the vitamin-exposed groups; limited number of animals; no control group exposed to air only; incidence not reported; administered material not well characterized	Morrison et al. (1981)
Male albino mouse of the Cobs strain (Charles River, CDR-1)	20 animals each in 3 treatment groups; 20 control animals	Exposure to glass fibres at 1070 fibres/cc (87% of fibres < 8 um in length and 100% < 3 um in diameter); styrene (300 ppm), fibrous glass (1070 fibres/cc) plus styrene (300 ppm), or filtered air (controls), 5 h/day, 5 days/week, for 6 weeks	No lung damage in animals exposed to fibrous glass alone; change in the cellularity of the bronchiolar lining where apocrine cells became predominant in animals exposed to glass fibres and styrene; this type of change found in 10% of styrene-exposed mice while most of the animals (90%) had thickened bronchiolar walls because of stratification of the bronchiolar epithelium; authors concluded that the glass fibres were biologically inert (possibly because	Moriset et al. (1979)

Table 17 (contd).

Male albino mouse of the Cobs strain (Charles River, CDR-1) (contd)			they were in non-pathogenic size range), but that the presence of such biologically inert respirable fibres enhances the toxic effects of styrene (possibly because of absorption of styrene on the fibre surfaces); limited number of animals and short exposure period	
Male Charles River/Sprague Dawley rat, albino male guinea-pig, and hamster	rats (46), guinea-pigs (32), hamsters (34)	Exposure to 0.7×10^6 fibres/litre (> 5 µm) glass fibres (0.4 mg/litre), 6 h/day, 5 days/week, for 90 days; animals sacrificed at 20, 50, or 90 days of exposure and at 6, 12, 18, or 24 months after exposure	Unlike the other fibrous dusts administered (asbestos, Fybex, and PKT), fibrous glass was not fibrogenic; some alveolar proteinosis, which had cleared after 2 years; authors concluded that fibrous glass satisfies criteria for a "biologically inert" dust; 18 months and 2 years after exposure, 2/19 exposed rats had bronchiolar adenomas compared with 0/19 control animals; however, numbers were too small to draw meaningful conclusions concerning carcinogenicity; study not well described; limited number of animals and short exposure and observation periods	Lee et al. (1981)

Table 17 (contd).

Species	Number	Protocol	Results and comments	Reference
SPF Fischer rat	2 exposed animals; 2 control animals	Exposure to glass fibre (JM 100), rock wool, resin-coated glass wool, uncoated glass wool, or air, at 10 mg/m^3 for 7 h/day, 5 days/week, over 50 weeks; animals sacrificed 4 months after inhalation; lung tissue examined by electron microscopy	All fibres produced variable degrees of focal fibrosis, type II alveolar cell proliferation, accumulation of cellular debris and lipid material and hyperplasia of both alveolar cells and cells lining the terminal airways; fibrosis less marked than for UICC chrysotile B; uncoated glass wool more reactive than coated variety; very small number of exposed and control animals	Johnson & Wagner (1980)
Rat and hamster (strains unspecified)	30 in each of the treatment groups; 20 controls for each species	Exposure to fibrous glass (uncoated, coated with phenol-formaldehyde resin, or starch binder) (average diameter, 0.5 µm; average length, 10 µm), at ~100 mg/m^3 6 h/day, 5 days/week, over 24 months; animals sacrificed at 6, 12, and 24 months; intratracheal injection of 1 - 10 doses of 3.5 mg of the same dusts in 150 rats and 60 hamsters	No discernable differences in tissue reactions for the 3 different types of fibrous glass; pulmonary response characterized by relatively small accumulation of macrophages without significant stromal change; authors conclude that fibrous glass "biologically inert"	Gross et al. (1970)
Guinea-pig and rat	100 guinea-pigs; 50 rats; no control group	Exposure to 5.05 - 5.16 mg/cm^3 (1.4 x 10^6 - 2.2 x 10^6 ppcf; mean diameter, 6 µm) for 20 months; guinea-pigs exposed for additional 20 months to 1.06 - 2.48 mg/cm^3 (mean diameter, 3 µm); animals sacrificed at periods of up to 40 months	Little evidence of dust reaction and no fibrosis	Schepers (1955); Schepers & Delahunt (1955)

Table 17 (contd).

Charles River CD Sprague Dawley derived male rat, albino male guinea-pig, and Syrian male hamster	Not reported	Exposure to 0.42 mg/litre (0.73 x 10^6 fibres/litre > 5 and < 10 µm; average diameter, 1.2 µm) glass fibre, 6 h/day, 5 days/week, for 90 days; animals sacrificed at periods of up to 2 years after exposure	At 90 days, macrophage reaction with alveolar proteinosis, which had disappeared 1 year after exposure; ferruginous bodies in hamsters and guinea-pigs; study not well described; short exposure and observation periods	Lee et al. (1979)
Inbred male BD-IX rat	Unspecified number in treatment and control groups	Repeated pulmonary lavage in animals exposed to 1340 fibres/cm^3 JM reference C102 glass fibres (diameter, 0.1 - 0.6 µm; length, 5 - 100 µm), 7 h/day, 5 days/week, for 6 months	Profound effects on alveolar macrophages (similar to those observed for crocidolite); however, temporal variation in the effects observed following pulmonary lavage (Sykes et al., 1983).	Miller (1980)
Male baboon (*Papio ursinus*)	10 animals in total; number in treated or control groups unspecified	Exposure to 7.54 mg/m^3 (1122 fibres/cc > 5 µm) of JM C102 and C104 fibrous glass 7 h/day, ~5 days/week, for up to 35 months; lung biopsy at intervals (material available for study at 8, 18, and 30 months and 6, 8, and 12 months after exposure)	Fibrosis was slightly more marked after 18 and 30 months exposure; also present in post-exposure biopsies; changes similar to those observed for UICC crocidolite (15.83 mg/m^3 - 1128 fibres/cc > 5 µm) but less severe; no evidence of malignancy; authors suggested that contrary results of others might be due to the diameter of the fibre and species of the exerimental animal; small number of animals and incidence in control group not reported; short exposure period in relation to life span of the animal	Goldstein et al. (1983, 1984)

Table 17 (contd).

Species	Number	Protocol	Results and comments	Reference
Albino rat of the Alderley Park (Wistar derived strain)	50 animals in each of the treatment groups; 50 controls	Exposure to 2.18 mg/m^3 (mean respirable dust) Saffil alumina fibres (aluminum oxide refractory fibre containing about 4% silica), 2.45 mg/m^3 thermally aged Saffil or air (median diameter, ~3 µm; median length, 10.5 - 62 µm) for 6 h/day, 5 days/week, for 86 weeks; animals sacrificed at 14, 27, or 53 weeks and at 85% mortality	Pulmonary reaction to both forms of Saffil was minimal (generally confined to the presence of groups of pigmented alveolar macrophages); no pulmonary neoplasms; authors concluded that Saffil may be regarded as a nuisance type dust, but acknowledged that the levels of respirable dust in the atmosphere were low	Pigott et al. (1981)
SPF Fischer rat	56 animals in each of the treatment groups; 56 controls	Exposure to rock wool without resin (227 fibres/cc; diameter < 3 µm; length, > 5 µm), glass wool without resin (323 fibres/cc; diameter < 3 µm; length > 5 µm), glass wool with resin (240 fibres/cc; diameter < 3 µm; length > 5 µm), glass microfibres (1436 fibres/cc; diameter < 3 µm; length > 5 µm), UICC Canadian chrysotile (3832 fibres/cc; diameter < 3 µm; length > 5 µm) at ~10 mg/m^3, or to air, for 7 h/day, 5 days/week for 12 months; animals sacrificed at 3, 12, or 24 months	Evidence of reaction to the dust in all exposed groups but less for MMMF than for chrysotile; greater reaction with chrysotile at 12 months than at 3 months, but little change for MMMF; in chrysotile exposed group, 12 lung neoplasms (11 adenocarcinomas and 1 adenoma) and broncho-alveolar hyperplasia (BAH) in 5 animals (No. = 48); glass microfibre - 1 pulmonary adenocarcinoma, BAH in 3 rats (No. = 48); rock wool - 2 adenomas, BAH in 1 rat (No. = 48); glass wool with resin - 1 pulmonary adenocarcinoma, BAH in 3 rats (No. = 48); glass wool without resin - 1 adenoma, BAH in 1 rat (No. = 47); controls - 0 neoplasms and BAH in 1 rat (No. = 48); incidence not specified	Wagner et al. (1984)

Table 17 (contd).

Fischer 344 rat	Group sizes at final sacrifice: 47 - 55 in each of the treatment groups; 48 - 53 controls	Exposure to Canadian chrysotile or glass microfibres (JM 100), at ~10 mg/m^3 for 7 h/day, 5 days/week, for 12 months; life-time observation with interim sacrifices at 3, 12, and 24 months; studies conducted in 2 laboratories (NIEHS and MRC)	At 3 months, minimal to mild cellular changes in both glass microfibres (gm)- and chrysotile (chrys)-exposed rats, which progressed througout the 12-month exposure period - changes less severe for gm- than for chrys; progression of changes in chrys-exposed rats but not in gm-exposed animals following cessation of exposure; increased incidence of pulmonary neoplasia found only in rats exposed to chrysotile	McConnell et al. (1984)
Male Syrian hamster, female Osborne-Mendel rat	Group sizes at final sacrifice: 47 - 60 hamsters in the treatment groups and 58 - 112 controls; 52 - 61 rats in each of the treatment groups and 59 - 125 controls	exposure to 300 fibres/cc (~0.3 mg/m^3), or 3000 fibres/cc (~3 mg/m^3) 0.45-µm mean diameter glass fibres (no binder), 100 fibres/cc (~10 mg/m^3) 3.1-µm mean diameter glass fibres (silicone lubricant), 10 fibres/cc (~1.2 mg/m^3), or 100 fibres/cc (~12 mg/m^3) 5.4-µm mean diameter glass fibres (binder-coated), 25 fibres/cc (~9 mg/m^3) 6.1-µm mean diameter glass fibres (binder-coated), 200 fibres/cc (~12 mg/m^3) 1.8-µm mean diameter refractory ceramic fibre (no binder), 200 fibres/cc (~10 mg/m^3)	Life span of hamsters exposed to glass fibres significantly longer than that of controls (with the exception of those exposed to 3.1-µm mean diameter fibres); foci of fibre-containing macrophages for both species exposed to glass fibres, but "little fibrosis" (peribronchiolar when observed); no tumours in MMMF-exposed animals with the exception of a mesothelioma in 1 hamster exposed to 1.8-µm mean diameter refractory ceramic fibre; asbestosis in crocidolite-exposed hamsters and rats; no pulmonary tumours in crocidolite-exposed hamsters	Smith et al. (1984, in press)

Table 17 (contd).

Species	Number	Protocol	Results and comments	Reference
Male Syrian hamster, female Osborne-Mendel rat (contd)		2.7 µm mean diameter mineral wool (no binder), 3000 fibres/cc (~7 mg/m^3) UICC crocidolite asbestos or air (sham and unmanipulated controls), nose-only, for 6 h/day, 5 days/week, for 24 months; life-time observation	(0/58), but 3/57 (5%) crocidolite-exposed rats developed lung neoplasms (compared with 0 in controls) (negative results in positive controls)	
SPF Wistar rat AF/HAN strain	48 exposed animals and 40 controls	Exposure to fibrous ceramic aluminum silicate glass at ~10 mg/m^3 (95 fibres/cc; diameter < 3 µm; length > 5 µm), for 7 h/day, 5 days/week, for 12 months; animals sacrificed at 12, 18, or 32 months	Survival of treated and control groups similar; interstitial fibrosis in animals exposed to ceramic fibres occurred to a lesser, but not significantly different, extent than that for chrysotile exposed animals; however, little peribronchiolar fibrosis in ceramic fibre-exposed rats; relatively large numbers of pulmonary neoplasms in rats exposed to ceramic fibre (tumours in 8 animals - 1 benign adenoma, 3 carcinomas, and 4 malignant histiocytomas) - no tumours in control animals; pattern of tumour development different from that for asbestos; large non-fibrous component of exposure aerosol	Davis et al. (1984)

Table 17 (contd).

SPF Fischer 344 rat	32 exposed animals and 32 controls	Exposure to glass microfibres at 50 000 fibres/cm³, for 5 - 6 h/day, for 2 or 5 days; animals sacrificed at 1, 4, 5, 8, and 12 months after exposure	No significant gross or histopathological changes	Pickrell et al. (1983)
Wistar IOPS AF/HAN rat	24 animals of each sex in each treatment group; 24 males and 24 females in control group	Exposure to (respirable dust fraction) glass wool (Saint Gobain) at 5 mg/m³ (73% < 20 um in length; 69% < 1 µm in diameter), rock wool (Saint Gobain) (40% < 10 µm in length, 49% < 2 µm in diameter), or glass microfibre (JM 100) (94% < 5 µm in length, 43% < 0.1 um in diameter), or Canadian chrysotile, for 5 h/day, 5 days/week, over 12 or 24 months; animals sacrificed at 0, 7, 12, and 16 months after 1-year exposure and 0 and 4 months after 2-year exposure	Simple alveolar macrophagic reaction with slight septal fibrosis (rock and glass wool); for glass microfibre, septal fibrosis slightly more marked; 1 pulmonary tumour (undifferentiated epidermoid carcinoma) in glass wool-exposed male (0 in controls) compared with 9 in chrysotile-exposed group; digestive tract tumours (unspecified) in 4 glass wool-exposed animals (0 in controls); small number of exposed animals and incidence not specified	Le Bouffant et al. (1984, in press)

Table 17 (contd).

SPF Fischer 344 rat	32 exposed animals and 32 controls	Exposure to glass microfibres at 50 000 fibres/cm^3, for 5 - 6 h/day, for 2 or 5 days; animals sacrificed at 1, 4, 5, 8, and 12 months after exposure	No significant gross or histopathological changes	Pickrell et al. (1983)
Wistar IOPS AF/HAN rat	24 animals of each sex in each treatment group; 24 males and 24 females in control group	Exposure to (respirable dust fraction) glass wool (Saint Gobain) at 5 mg/m^3 (73% < 20 µm in length; 69% < 1 µm in diameter), rock wool (Saint Gobain) (40% < 10 µm in length, 49% < 2 µm in diameter), or glass microfibre (JM 100) (94% < 5 µm in length, 43% < 0.1 µm in diameter), or Canadian chrysotile, for 5 h/day, 5 days/week, over 12 or 24 months; animals sacrificed at 0, 7, 12, and 16 months after 1-year exposure and 0 and 4 months after 2-year exposure	Simple alveolar macrophagic reaction with slight septal fibrosis (rock and glass wool); for glass microfibre, septal fibrosis slightly more marked; 1 pulmonary tumour (undifferentiated epidermoid carcinoma) in glass wool-exposed male (0 in controls) compared with 9 in chrysotile-exposed group; digestive tract tumours (unspecified) in 4 glass wool-exposed animals (0 in controls); small number of exposed animals and incidence not specified	Le Bouffant et al. (1984, in press)

7.1.1.1 Fibrosis

No evidence of fibrosis of the lung was observed in most of the inhalation studies conducted to date on the mouse, rat, guinea-pig, and hamster, exposed to concentrations of glass fibres of up to 100 mg/m^3 for periods ranging from 2 days to 24 months (Schepers & Delahunt, 1955; Gross et al., 1970; Lee et al., 1979; Morriset et al., 1979; Lee et al., 1981; Pickrell et al., 1983; Smith et al., 1984). In general, the tissue response in these studies was confined to the accumulation of pulmonary macrophages. However, reversible alveolar proteinosis was reported in 2 rather limited studies by the same investigators in which several species were exposed to relatively high concentrations of glass fibres (0.4 mg/litre) (Lee et al., 1979, 1981). In the more extensive investigations of McConnell et al. (1984) and Wagner et al. (1984), in which rats were exposed to glass microfibres at 10 mg/m^3 for 12 months, "minimal interstitial cellular reactions with no evidence of fibrosis" were reported.

However, in contrast, Johnson & Wagner (1980) reported that electron microscopic examination of the lung tissue of 2 rats exposed to glass fibre, rock wool, resin-coated glass wool, or uncoated glass wool at 10 mg/m^3 for 50 weeks, revealed focal fibrosis. In a limited study, Le Bouffant et al. (in press) reported slight septal fibrosis that was a little more marked in rats exposed to glass microfibre than in those exposed to rock and glass wool. Fibrosis was also observed in baboons exposed to 7.54 mg/m^3 glass fibres for 35 months (Goldstein et al., 1983). However, the incidence and severity in control animals were not reported. Moreover, the fibrotic lesions observed in this study were similar to those caused by lung mites (*Pneumonyssus* sp.), commonly present in baboons from this geographical area (McConnell et al., 1974). In a more recent study in which cynomolgus monkeys were exposed for 18 months to up to 15 mg glass fibres/m^3 with mean diameters of 0.5 - 6 μm, the tissue response was confined to accumulation of pulmonary macrophages and granulomas (Mitchell et al., 1986).

In all cases, the tissue reactions in animals exposed to fibrous glass or glass wool were much less severe than those produced by exposure to equal masses of chrysotile or crocidolite (Johnson & Wagner, 1980; Lee et al., 1981; Goldstein et al., 1983; McConnell et al., 1984; Smith et al., 1984; Wagner et al., 1984). Moreover, in contrast to fibrotic changes resulting from exposure to asbestos, tissue responses did not progress following cessation of exposure to these MMMF (McConnell et al., 1984; Wagner et al., 1984).

Results concerning the effects of fibre coating on the potential of fibrous glass or glass wool to cause lung-tissue damage are limited and contradictory. Whereas Gross et al.

(1970) concluded that there were no discernible differences in the tissue reactions for uncoated fibrous glass compared with those for fibrous glass coated with phenol-formaldehyde resin or starch binder, Johnson & Wagner (1980) concluded that uncoated glass wool was more reactive than resin-coated varieties.

Pigott et al. (1981) reported minimal pulmonary reaction (generally confined to the presence of groups of pigmented alveolar macrophages) in rats exposed for 86 weeks to relatively low concentrations of normal or thermally aged aluminium oxide refractory fibres containing about 4% silica (mean respirable dust concentrations of 2.18 and 2.45 mg/m^3, respectively; median fibre diameter, 3.3 μm). However, it should be noted that the median fibre diameter of the material is relatively large. Davis et al. (1984) reported that interstitial fibrosis occurred to a lesser extent in rats, exposed for 12 months to fibrous ceramic aluminium silicate glass at 10 mg/m^3 (with a rather large non-fibrous component in the aerosol), than in chrysotile-exposed animals. In contrast, Smith et al. (1984, in press) did not find any fibrosis in hamsters and rats exposed to 12 mg/m^3 refractory ceramic fibres (with about twice the airborne fibre concentration) for a period of 2 years.

With respect to the fibrogenic effects of combined exposure to MMMF and other airborne pollutants, Morriset et al. (1979) concluded that, though glass fibres appeared to be biologically inert in their study (possibly because they were in a nonpathogenic size range), the presence of these fibres enhanced the toxic effects of styrene in mice (possibly because of the absorption of styrene on the fibre surface).

The discrepancies in the results of different investigations concerning the severity of tissue response following inhalation of glass and ceramic fibres may be the result of variation in fibre size distribution. For example, Johnson & Wagner (1980) suggested that the fibrotic changes observed in their study might have been due to the fact that their samples contained a greater proportion of fibres longer than 5 μm than those of other investigators. However, it is difficult to draw general conclusions in this regard because of inconsistencies in the characterization of the airborne fibre size distributions and in doses administered in different studies.

7.1.1.2 Carcinogenicity

In none of the inhalation studies conducted to date has there been a statistically significant excess of lung tumours in animals exposed to glass fibres (including glass microfibres) or rock wool. However, there have only been a few relevant studies (Gross et al., 1970, 1976; Lee et al., 1981; McConnell et al., 1984; Wagner et al., 1984; Le Bouffant et al., in press; Mühle et al., in press; Smith et al., in press), and group sizes at termination in several of these investigations were small by current standards. However, in several of the relevant studies, there were small, but not statistically significant, increases in tumour incidence in exposed animals. For example, 2/19 (10.5%) Sprague Dawley rats exposed for only 90 days to fibrous glass at 0.4 mg/litre had bronchiolar adenomas 2 years after exposure, compared with none in the control group (Lee et al., 1981). Similarly, there were 1 - 2 neoplasms (2.1 - 4.2%) in each group of Fischer rats exposed to rock wool (with or without resin), glass wool, or glass microfibre at 10 mg/m^3 for 12 months, compared with none in the controls (Wagner et al., 1984). No mesotheliomas have been observed in animals exposed to glass fibres or rock wool by inhalation; moreover, in most of the carcinogenicity bioassays conducted to date, similar mass concentrations of chrysotile asbestos have been far more potent in inducing lung tumours than these MMMF (Lee et al., 1981; McConnell et al., 1984; Wagner et al., 1984; Smith et al., in press). However, data are not sufficient to draw conclusions concerning the relative potency of various fibre types, because the concentration of respirable fibres was not reported in many of these studies.

Morrison et al. (1981) reported "neoplastic, hyperplastic, and metaplastic lesions in the bronchi and near the bronchio-alveolar junctions of the lung" in 12 A strain mice sacrificed 90 days after exposure to 300 mg "crushed fiberglass insulation" (80% of fibres 2.5 μm wide) in bedding, every 3 days for 30 days. However, it is difficult to assess the results of this study, since there was no control group exposed to air only. In addition, only the total number of lesions, rather than the incidence, was specified and the administered material was not well characterized. Furthermore, these results seem inconsistent with those of other investigators, particularly in view of the relatively short exposure period and the small number of exposed animals.

Pigott et al. (1981) reported the absence of pulmonary neoplasms in 50 Wistar-derived Alderley Park rats exposed for 86 weeks to relatively low mean respirable dust concentrations of normal or thermally aged aluminium oxide refractory fibre containing about 4% silica at 2.18 and 2.45 mg/m^3, respectively. Davis et al. (1984) reported "relatively large numbers of pulmonary neoplasms" in Wistar rats of the Han strain exposed for 12 months to fibrous ceramic aluminium silicate glass at 10 mg/m^3. Pulmonary neoplasms (3 carcinomas and 1 adenoma) developed in 8/48 exposed animals (16.6%). However, in 4 of the animals, the tumours were malignant histiocytomas; these neoplasms have not generally been associated with asbestos exposure. There was also one peritoneal mesothelioma in another animal in the exposed group (the non-fibrous content of the exposure aerosol in this study was relatively large).

In Syrian golden hamsters exposed to approximately 12 mg "refractory ceramic fibres"/m^3 (200 fibres/cc) for 2 years, there was a single malignant mesothelioma. However, no other primary lung tumours were observed in 51 animals surviving the exposure period, and no primary lung tumours were observed in similarly exposed Osborne-Mendel rats. It should be noted, though, that only 3 of the 57 rats and none of the 58 hamsters that survived the exposure period following inhalation of 3000 fibres/cm^3 of crocidolite for 2 years developed lung tumours in this study (Smith et al., in press).

7.1.2 Intratracheal injection

Administration by the intratracheal route does not simulate the exposure of man, since an uneven dose of fibres is artificially deposited in the respiratory tract. For example, the production of fibrous granulomas in such studies may be a function of "bolus" events and might not be observed in inhalation studies where fibre deposition is more diffuse. Although this limitation must be borne in mind in interpreting the results of intratracheal studies, such investigations have confirmed the effects of MMMF observed following inhalation and have provided additional information on the importance of fibre size in the pathogenesis of disease. The results of available studies involving intratracheal administration of MMMF are presented in Table 18. For example, following intratracheal administration of uncoated or coated glass fibres (1 - 10 doses of 3.5 mg) to rats, Gross et al. (1970) observed reversible polyploid proliferative lesions, but no alveolar fibrosis. For hamsters, the tissue reaction was similar with the exception that diffuse, bland, acellular collagenous pleural fibrosis was also observed. Pickrell et al. (1983) observed proliferation of bronchioalveolar epithelial cells, chronic inflammation of the terminal bronchioles and alveolar ducts, and slight pulmonary fibrosis following intratracheal administration of glass microfibres to Syrian hamsters (total dose, 2 mg; count median diameter, 0.2 μm). Fibrosis was not observed in animals exposed in a similar manner to 2 instillations of 3 types of glass fibre "household insulation" (total dose, 17 - 21 mg; count median diameters, 2.3 - 4.1 μm). A dose-related trend was reported by Renne et al. (1985) in the incidence of alveolar septal fibrosis in hamsters exposed by the intratracheal route to 15 weekly doses (0.05 - 10 mg) of fibrous glass with a median fibre diameter and length of 0.75 and 4.30 μm, respectively. The authors reported that the pulmonary response to quartz alone and to quartz in combination with ferric oxide was considerably more severe than that to fibrous glass. Drew et al. (in press) reported a granulomatous foreign body response following intra-tracheal instillation of a single dose of 20 mg of "long"

Table 18. Intratracheal injection studies

Species	Number in groups	Protocol	Results and comments	Reference
Rat, hamster	12 - 30 animals per test group; 20 controls of each species	Injection (1 - 10 doses) of 3.5 mg uncoated or coated (phenol-formaldehyde resin or starch binder) "glass fibres" (length $\leq$ 50 µm)	Rats: reversible polyploid proliferative lesions but but no alveolar fibrosis; hamsters: tissue reaction similar to that observed in rats, except that diffuse, bland acellular collagenous pleural fibrosis also seen; observation period not reported; controls not held under same conditions as test group	Gross et al. (1970)
Male Syrian hamster (Charles River Sch:(SYR))	20 animals per test group; 30 controls	Instillation of (2 doses) of 2 - 21 mg of one of 2 types of "bare glass" or 3 types of "household insulation"; count median diameters ranged from 0.1 to 4.1 µm; animals sacrificed at 1, 3.5, and 11 months	Glass microfibres produced proliferation of bronchio-alveolar epithelial cells, chronic inflammation of the terminal bronchioles and alveolar ducts, and slight pulmonary fibrosis; no fibrosis after instillation of the "household insulation"	Pickrell et al. (1983)
Male Syrian golden hamster Lak:LVG	25 animals each in test, saline, and cage control groups	Instillation (15 weekly doses of 0.05 - 10 mg fibrous glass with mean diameter and length of 0.75 and 4.30 µm, respectively; animals maintained until 28% survival in each group or until 24.5 months of age	Dose-related trend in the incidence of alveolar septal fibrosis; the pulmonary response to quartz and ferric oxide was considerably more severe	Renne et al. (1985)

6

Table 18 (contd).

Species	Number in groups	Protocol	Results and comments	Reference
SPF Fischer-344 male rat	50 animals per test and control groups	Instillation of 20 mg of short (nominal diameter and length, 1.5 x 5 µm) and long (1.5 x 60 µm) glass fibres or 10 weekly instillations of 0.5 mg glass fibres	A single instillation of short or long fibres produced a granulomatous foreign body response, whereas weekly instillations elicited only a macrophage response; severe pulmonary response with fibrosis after exposure to crocidolite	Drew et al. (in press)
Guinea-pig	5 - 8 animals per test and control groups	Instillation of 0.5 cc of 0.5% (1 group) of 5% (2 groups) glass fibres with mean diameters of 6, 3, and 1 µm	Severity of the tissue reaction (no fibrosis) inversely proportional to the fibre diameter	Schepers & Delahunt (1955)
Guinea-pig	30 animals per test and control groups	Instillation (2 - 8 doses) of "short" (< 5 µm) and "long" (> 10 µm) glass fibres; total doses of 12 and 25 mg, respectively	Tissue reaction varied with fibre length, with fibrosis resulting only following instillation of "1 mg"	Kuschner & Wright (1976); Wright & Kuschner (1977)
Male Syrian golden hamster	136 or 138 animals per test group	8 weekly instillations of 1 mg JM 104 (wet milled) glass fibre, 50% with diameters < 3 µm; animals observed up to week 130	Incidence of lung carcinomas was slightly less than that of animals exposed to UICC crocidolite; however, the incidence of sarcomas (thorax) and mesotheliomas was greater in glass fibre-exposed animals; not confirmed in investigations in other laboratories	Mohr et al. (1984)

Table 18 (contd).

Female Wistar rat (Wistar-W11/KiBlegg)	34 animals per test and control groups	Instillation of 0.05 mg (20 doses) of JM 104/Tempstran 475 glass fibres (50% with lengths and diameters < 3.2 µm and 0.18 µm, respectively)	Lung tumours in 5/34 animals (1 adenoma), 2 adenocarcinomas, and 2 squamous cell carcinomas); 9 lung carcinomas and 8 mesotheliomas in 142 animals with similar exposure to crocidolite (length < 2.1 µm; diameter < 0.2 µm for 50% of fibres)	Pott et al. (1987)
Syrian golden hamster	35 animals of each sex per test and control groups	Instillation of 1 mg, every 2 weeks, for 52 weeks, of 1 glass fibres (88% with diameters < 1 µm; 58% with lengths < 5 µm); animals observed for up to week 85	"Glass fibre granulomas", but no tumours and no indication that the glass fibres enhanced respiratory tract tumours induced by benzo(a)pyrene (26 doses of 1 mg glass fibres and 1 mg benzo(a)pyrene); no tumours observed in positive (crocidolite-exposed) control group	Reuzel et al. (1983); Feron et al. (1985)

Table 18 (contd).

Species	Number in groups	Protocol	Results and comments	Reference
Osborne-Mendel rat and Syrian golden hamster	22 - 25 animals in test groups at final autopsy; 24 - 25 saline controls and 112 - 125 cage controls at final autopsy	Instillation of 2 mg of either glass fibre (mean diameter, 0.45 µm) or 1.8-µm mean diameter refractory ceramic fibre, once a week, for 5 weeks; life-time observation	For refractory ceramic fibre, significant reduction in the life span of hamsters (479 days compared with 567 days in vehicle controls) and a chronic inflammatory response in the lungs of rats (probably due to deposition of large quantities of foreign materials); fibrosis in rats exposed to 0.45-µm mean diameter glass fibres; no pulmonary tumours in hamsters or rats exposed to either fibre; pulmonary tumours in 8% (rats) and 7.4% (hamsters) exposed to crocidolite	Smith et al. (in press)

(nominal diameter and length, 1.5 μm x 60 μm) and "short" (1.5 μm x 5 μm) glass fibres in rats, which they attributed to the administration technique. However, following 10 weekly instillations of 0.5 mg, only a macrophage response was elicited. In contrast, a much more severe pulmonary response accompanied by fibrosis resulted from exposure to crocidolite.

In a study conducted as early as 1955, it was reported that the severity of the tissue reaction (no fibrosis) following intratracheal administration of 3 samples of glass fibres (three 0.5-cc doses of 0.5% (1 group) or 5% (2 groups) solutions) was inversely proportional to fibre diameter (mean values, 6 μm, 3 μm, and 1 μm) (Schepers & Delahunt, 1955). Kuschner & Wright (1976) observed that the severity of tissue reaction following intratracheal administration of glass fibres (2 - 8 instillations of 12 - 25 mg) to guinea-pigs varied with fibre length, with fibrosis resulting only from exposure to samples containing fibres mainly longer than 10 μm (Wright & Kuschner, 1977).

Mohr et al. (1984) found that the incidence of lung carcinomas in Syrian golden hamsters following a rather unusual intratracheal administration technique involving weekly instillations of relatively low doses of fibres (8 weekly doses of 1 mg each of wet-milled JM 104 glass fibres, 50% with diameters < 0.3 μm) was slightly less than that for animals exposed to UICC crocidolite. Surprisingly, the incidence of sarcomas (thorax) and mesotheliomas was greater in the glass fibre-exposed animals. It has been suggested by the authors that these results might be attributable to the relatively short fibre lengths of the UICC crocidolite (Pott et al., 1987). In a more recent but similar study in the same laboratory, lung tumours (1 adenoma, 2 adenocarcinomas, 2 squamous cell carcinomas) were observed in 5 out of 34 female Wistar rats receiving 20 weekly applications of 0.05 mg each of JM 104/Tempstran 475 glass fibres (50% of fibre lengths < 3.2 μm; 50% of fibre diameters < 0.18 μm) (Pott et al., 1987). In rats exposed to similar doses of crocidolite on the same schedule (50% of fibre lengths < 2.1 μm; 50% of fibre diameters < 0.20 μm), lung carcinomas were observed in 9 and mesotheliomas in 8 out of a total of 142 animals examined.

These results concerning the carcinogenicity of intratracheally administered glass fibres have not been confirmed in other laboratories. For example, Reuzel et al. (1983) and Feron et al. (1985) reported "glass fibre granulomas", but no tumours, in hamsters receiving intratracheal instillations of 1 mg glass fibres (88% of fibre diameters < 1 μm and 42% > 5 μm length), once every 2 weeks, for 1 year; moreover, there was no indication that glass fibres enhanced the development of respiratory tract tumours induced by benzo(a)pyrene (26 doses of 1 mg glass fibres and 1 mg benzo(a)pyrene). However, it should be noted that no tumours were observed in the positive controls (i.e., the crocidolite-exposed group). The strain of animals and the form of glass fibre administered in this study were similar to those used by Mohr et al. (1984). Feron et al. (1985) suggested that the discrepancies between the results of the 2 investigations might be attributable to the differences in the length of the observation periods (85 weeks compared with 113 weeks in the Mohr et al. (1984) study) or to the effects of prolonged repeated dosing (i.e., disturbance of fibres by acute pulmonary reaction after each of the 26 doses). Kuschner (in press) suggested that the difference might be due to different instillation techniques.

In an additional study by Smith et al. (in press), the mean life span of 25 hamsters receiving 2 mg of 1.8-μm mean diameter refractory ceramic fibre, intratracheally, once a week, for 5 weeks, was significantly reduced (479 days compared with 567 days in vehicle controls). However, no pulmonary tumours were observed in the hamsters or in rats receiving either 0.45-μm mean diameter glass fibres or 1.8-μm mean diameter refractory ceramic fibre, on a similar schedule. Two out of 25 rats (8%) and 20 out of 27 hamsters (74%) similarly exposed to crocidolite developed pulmonary tumours. Pulmonary response to the various types of MMMF in this investigation was restricted to a chronic inflammatory reaction in rats exposed to the ceramic fibre (believed to be due to deposition of large quantities of foreign materials in the lung) and fibrosis in the same species exposed to glass fibres with a mean diameter of 0.45 μm.

7.1.3 Intrapleural, intrathoracic, and intraperitoneal administration

Although introduction of mineral fibres into the pleura and peritoneum of animal species does not simulate the route of exposure of man, such studies have made it possible to clarify a number of questions that could not feasibly have been investigated using the inhalation model. Studies involving intrapleural or intraperitoneal injection also serve as useful screening tests to develop priorities for further investigation. Both have been used to assess the potential for fibres to induce mesothelioma, when placed in contact with a target tissue. In addition, the fibrogenic potential of both particles and fibres has been examined in studies involving intraperitoneal injection and short observation periods. However, factors that affect fibre deposition and translocation, and defence mechanisms that determine retention of fibres within the lung are not taken into consideration in this experimental model and these disadvantages must always be borne in mind in interpreting the results of such studies.

Both fibrosis and malignant tumours have resulted following the implantation of various types of MMMF into the pleural, thoracic, or peritoneal cavities of various species; the results of the relevant studies are summarized in Table 19. These investigations have been most important in focusing attention on the role of fibre size and shape in the induction of disease. For example, in studies by Davis (1972, 1976), intrapleural injection of long-fibred samples of fibrous glass produced massive fibrosis, while short-fibred samples produced only discrete granulomas with minimal fibrosis. In 1972, on the basis of their study involving intrapleural implantation of 17 fibrous materials (including 6 types of fibrous glass) in rats, Stanton & Wrench (1972) first hypothesized that "the simplest incriminating feature for both carcinogenicity and fibrogenicity seems to be a durable fibrous shape, perhaps in a narrow range of size". In subsequent studies by the same authors, involving intrapleural implantation of numerous dusts including fibrous glass, it was found that the probability of development of pleural sarcomas was best correlated with the number of fibres with diameters of less than 0.25 μm and lengths greater than 8 μm (Stanton et al., 1977, 1981; Stanton & Layard, 1978). However, probabilities were also "relatively" high for fibres with diameters < 1.5 μm and lengths > 4 μm. On the basis of an extensive series of studies involving intraperitoneal administration to rats of asbestos, fine glass fibres, and nemalite, Pott (1978) hypothesized a model in which the carcinogenic potency of fibres is a continuous function of length and diameter. Based on the results of further studies, which indicated that tumour incidence for long, thin, nondurable fibres was less than that for durable fibres in the same size range, Pott et al. (1984) concluded that carcinogenicity is also a function of the durability of fibres in the body.

In general, chrysotile has been more potent than equal masses of glass fibres in inducing tumours following intrapleural injection (Monchaux et al., 1981; Wagner et al., 1984). However, the potency of glass fibres varies markedly as a function of fibre size distribution; intraperitoneal administration of continuous glass filament with mean diameters of 3, 5, and 7 μm did not increase tumour incidence (Pott et al., 1987). On the basis of studies involving intrapleural injection of rock-, slag-, or glass wool or glass microfibres, it also appears that tumour incidence is roughly proportional to the number of fibres injected (Wagner et al., 1984). In a recent study, in which actinolite and a basalt wool that contained equal numbers of fibres (length ≥ 5 μm) were administered to rats, tumour incidence was similar, though the diameters differed by an order of magnitude (0.1 μm versus 1.1 μm) (Pott et al., in press). However, it should be noted that the fibre length distributions were substantially different as was the mass injected.

Table 19. Intrapleural, intrathoracic, and intraperitoneal administration

Species	Number in groups	Protocol	Results and comments	Reference
SPF Sprague Dawley rat	8 groups of 48 animals each; 24 controls	Intrapleural injection of 20 mg respirable samples of uncoated and resin-coated Swedish rock wool, German slag wool, and English glass wool, American glass microfibres (JM 100), and chrysotile; number of fibres > 5 µm ranged from 1.2×10^8 to 4.2×10^8; value for chrysotile, 196×10^8	Mesotheliomas in 6 (12.5%) of chrysotile-exposed rats, 4 (8.3%) exposed to glass fibres, 3 (6.25%) exposed to rock wool with resin, 2 (4.2%) exposed to rock wool without resin, 1 (2.1%) exposed to glass wool, and none exposed to slag wool; tumour incidence roughly proportional to number of fibres injected	Wagner et al. (1984)
Wistar and Sprague Dawley rat	~50 animals; 50 controls	Intraperitoneal administration of 2 - 10 mg different preparations of glass microfibres (JM 100, 104), basalt wool, rock wool, and slag wool, and chemically treated (HCl or NaOH) glass microfibres (JM 104)	Tumour incidence varied from none (rock wool) to 73% (JM 104, 1-h ball milling); carcinogenicity reduced by HCl and NaOH	Pott et al. (1984)
SPF Wistar rat of the AF/HAN strain	32 animals in exposed group	Intraperitoneal injection of 25 mg fibrous ceramic aluminium silicate glass	Tumours in 3 (9.3%) animals (compared with 90% in chrysotile-exposed animals); first tumour occurred ~850 days after injection compared with 200 days for chrysotile (contrasts with the results of inhalation study); no concurrent control group	Davis et al. (1984)

Table 19 (contd).

Sprague Dawley rat	10 groups of 10 animals inhaling radon and receiving intrapleural injection of various mineral dusts; 60 controls exposed to radon only	Exposure to 3000 WL radon/day, for 10 h/day, 4 days/week, over 10 weeks; 2 weeks later, intrapleural injection of 2 mg of one of 10 mineral dusts (including JM 104 glass fibres); life-span observation	Proportion of lung cancer in radon only exposed rats was 28% compared with 68% in group with combined exposure to radon and intrapleural injection of mineral dust (tumour type also varied); in glass fibre-exposed group, no mesotheliomas but tumours in 6 (60%) animals; number in each group too small to allow comparison of effect by fibre type	Bignon et al. (1983)
Balb/C mouse and rat (strain unspecified)	18 - 25 animals in exposed group	Intrapleural injection of 10 mg of 4 samples of boron silicate glass fibre (long- or short-fibred samples with mean diameters of 0.05 or 3.5 µm); animals killed 2 weeks - 18 months after injection; intraperitoneal injection of 10 mg long-fibred, 0.05-µm diameter sample in both mice and rats; life-span observation	Long-fibred samples produced massive fibrosis, while short fibred samples produced only discrete granulomas with minimal fibrosis; in first study, no tumours in 37 mice sacrificed at 18 months; in second study, 3 tumours in 18 rats and and 3 tumours in 25 mice; no control group and observation period in first study short	Davis (1976)
SPF Male Sprague Dawley rat	45 exposed; 32 controls	Intrapleural injection of 20 mg glass fibre (JM 104; mean length, 5.89 µm; mean diameter, 0.229 µm); life-span observation	Survival times for animals injected with glass fibres longer than for controls; tumours in 13% of animals (similar to that for leached (44 - 64% Mg removed) chrysotile	Lafuma et al. (1980); Monchaux et al. (1981)

Table 19 (contd).

Sprague Dawley rat	10 groups of 10 animals inhaling radon and receiving intrapleural injection of various mineral dusts; 60 controls exposed to radon only	Exposure to 3000 WL radon/day, for 10 h/day, 4 days/week, over 10 weeks; 2 weeks later, intrapleural injection of 2 mg of one of 10 mineral dusts (including JM 104 glass fibres); life-span observation	Proportion of lung cancer in radon only exposed rats was 28% compared with 68% in group with combined exposure to radon and intrapleural injection of mineral dust (tumour type also varied); in glass fibre-exposed group, no mesotheliomas but tumours in 6 (60%) animals; number in each group too small to allow comparison of effect by fibre type	Bignon et al. (1983)
Balb/C mouse and rat (strain unspecified)	18 - 25 animals in exposed group	Intrapleural injection of 10 mg of 4 samples of boron silicate glass fibre (long- or short-fibred samples with mean diameters of 0.05 or 3.5 µm); animals killed 2 weeks - 18 months after injection; intraperitoneal injection of 10 mg long-fibred, 0.05-µm diameter sample in both mice and rats; life-span observation	Long-fibred samples produced massive fibrosis, while short fibred samples produced only discrete granulomas with minimal fibrosis; in first study, no tumours in 37 mice sacrificed at 18 months; in second study, 3 tumours in 18 rats and and 3 tumours in 25 mice; no control group and observation period in first study short	Davis (1976)
SPF Male Sprague Dawley rat	45 exposed; 32 controls	Intrapleural injection of 20 mg glass fibre (JM 104; mean length, 5.89 µm; mean diameter, 0.229 µm); life-span observation	Survival times for animals injected with glass fibres longer than for controls; tumours in 13% of animals (similar to that for leached (44 - 64% Mg removed) chrysotile	Lafuma et al. (1980); Monchaux et al. (1981)

Table 19 (contd).

Syrian golden hamster	60 males in each exposed group; 60 controls	Intrapleural injection of 25 mg of 6 samples of glass fibres	Tumours in 15% for sample with mean diameter 0.1 µm and 82% longer than 20 µm; 3.3% incidence for sample with 0.33 µm mean diameter and 46% longer than 20 µm; similar incidence for sample with 1.23 µm mean diameter and 34% longer than 20 µm; no tumours in hamsters treated with fibres with mean diameters of 0.09, 0.41, or 0.49 µm, but with only 0 - 2% of fibres longer than 10 µm	Smith et al. (1980)
Balb/C mouse	25 in each exposed group	Intrapleural injection of 10 mg of several fibrous dusts including 2 samples of boron silicate glass fibre (mean diameters, 0.05 - 1 µm and 2.5 - 4 µm; length < 10 µm) and three "man-made insulation fibres" (alumino-silicate: mean diameter, 4 µm; calcium-silicate: mean diameter, 10 µm; calcium-alumino-silicate: mean diameter, 4 µm); animals killed 2 weeks - 18 months after exposure	Long-fibred dust specimens produced widespread cellular granulomata, which formed firm adhesions, gradually replaced by fibrous tissue; short-fibred dust specimens produced smaller granulomata without adhesions; non-fibrous mineral rocks, when finely ground, also produced small non-adherent granulomata; final degree of fibrosis correlated with initial cellularity of lesions; no mention of control group	Davis (1972)

Table 19 (contd).

Species	Number in groups	Protocol	Results and comments	Reference
Female Wistar albino rat	10 in each group	Intraperitoneal injection of 50 mg of several fibrous dusts including "long" (mean length, 12.9 µm) and "short" (mean length, 2.4 µm) fibrous glass; animals killed at periods of up to 330 days after exposure	Initial foreign body reaction followed by fibroblast infiltration, fibrosis, and a few areas of reactive mesothelium, but no malignant transformation; tissue response of long and short fibres similar; authors concluded that "mechanical irritation does not contribute to the induction of mesotheliomas"	Englebrecht & Burger (1975)
SPF Wistar rat	up to 36 in each of exposed groups	Intrapleural injections of 20 mg ceramic fibres, fibreglass, glass powder, aluminium oxide, and 2 samples of SFA chrysotile; lifetime observation	Occasional mesotheliomas for ceramic fibre (3) and glass powder (1); none for glass fibres compared with (23) and (21) for the SFA chrysotile; estimated carcinogenicity factors ($\times 10^9$) were: for ceramic fibres, 0.16; for glass powder, 0.04; for SFA chrysotile, 2.28 - 2.85; authors concluded that results were consistent with the hypothesis that finer fibres are more carcinogenic	Wagner et al. (1973)

Table 19 (contd).

Female SPF Osborne-Mendel rat	30 - 50 rats in test and control groups	Intrathoracic implantation on a fibrous glass vehicle of 40 mg of 72 samples of 12 different fibrous materials (including 22 samples of fibrous glass); 2-year observation period	Probability of pleural sarcoma best correlated with the number of fibres with a diameter < 0.25 µm and length > 8 um, but relatively high correlations also noted for fibres with a diameter up to 1.5 µm and length > 4 µm; morphological observations indicated that short fibres and large diameter fibres were inactivated by phagocytosis and that negligible phagocytosis of long, thin fibres occurred; authors concluded that "the simplest incriminating feature for both carcinogenicity and fibrogenicity seems to be a durable fibrous shape, perhaps in a narrow range of size"	Stanton & Wrench (1972); Stanton et al. (1977, 1981); Stanton & Layard (1978)
SPF albino Wistar rat (Alderley Park strain)	40 animals in exposed and control groups	Intraperitoneal injection of 0.2 ml of a 10% suspension of Saffil alumina (median diameter, 3.6 µm; median length, 17 µm) or Saffil zirconia (median diameter, 2.5 µm; median length, 11 µm); animals killed 6 months after dosing	Nodular deposits of connective tissue; no fibrosis	Styles & Wilson (1976)

Table 19 (contd).

Species	Number in groups	Protocol	Results and comments	Reference
Female Wistar rat	30 - 32 animals in exposed and control groups	Intraperitoneal injection of 0.5 mg glass fibre (JM 104; 90% of fibre lengths < 8.2 µm, 90% of fibre diameters < 0.42 µm), 0.5 mg crocidolite (90% of fibre lengths < 7.7 µm, 90% of fibre diameters < 0.36 µm), 0.5 mg Calidria chrysotile, 1 mg UICC Canadian chrysotile B (90% of fibre lengths < 3.6 µm, 90% of fibre diameters < 0.18 µm)	Significant increases in tumour incidence for glass fibre (17%), crocidolite (55%), and Canadian chrysotile (84%); incidence in Calidria chrysotile-exposed group similar to controls (6%)	Mühle et al. (in press)
Male Syrian hamster and female Osborne-Mendel rat	~25 animals per test and control groups at necropsy	Intraperitoneal injection of 25 mg of 0.45-µm mean diameter fibrous glass, 1.8-µm mean diameter refractory ceramic fibre or crocidolite (fibre size distributions similar to those presented in Table 17); life-span observation	Abdominal mesotheliomas in 32% of rats exposed to glass fibre, 83% in rats exposed to refractory ceramic fibre, 80% in rats exposed to crocidolite (0 in controls); abdominal mesotheliomas in 13 and 24% of hamsters exposed to refractory ceramic fibre, 40% in hamsters exposed to crocidolite (0 in controls)	Smith et al. (in press)

Table 19 (contd).

Female Sprague Dawley and Wistar rat	number examined in each group ranged from 30 to 99	Intraperitoneal administration (single or up to 5 repeated doses); total administered material of treated or untreated samples of JM 100 (2 - 10 mg) or JM 104 (5 - 10 mg) glass microfibres, JM 106 (10 mg), slag wool (Rheinstahl; Zimmerman) (40 mg), Swedish rock wool (75 mg), ES5 (10 - 40 mg) and ES7 (40 mg) glass filaments, and basalt wool (Grünzweig & Hartman) (75 mg); intraperitoneal administration of ES3 (50 - 250 mg) or ES5 (250 mg) glass filaments by laparotomy	Potency to induce tumours related principally to size distribution of the fibres and to durability; tumour incidence for MMMF varied from 2.1 to 87% (the latter incidence occurred in the study using 5 mg NaOH-treated JM 104 glass microfibres); no increase in tumour incidence with glass filaments with diameters of 3, 5, and 7 µm	Pott et al. (1987)

Table 19 (contd).

Species	Number in groups	Protocol	Results and comments	Reference
Female Wistar rat	35 - 50 per group; 100 controls	Intraperitoneal administration (single or 5 repeated doses) of 0.01 - 0.25 mg actinolite, 0.05 - 1 mg chrysotile, 5 mg glass fibre JM 104/475, 75 mg basalt wool, 45 mg ceramic wool (Fibrefrax), 75 mg ceramic wool (Manville)	Fibre number (length > 5 µm; diameter < 3 µm) injected versus tumour rate reported: actinolite, 100×10^6 fibres (56% tumours); chrysotile, 200×10^6 fibres (68% tumours); JM 104/475, 680×10^6 fibres (64% tumours); basalt wool, 60×10^6 fibres (57% tumours); ceramic wool (Fiberfrax), 150×10^6 fibres (70% tumours); ceramic wool (Manville), 21×10^6 fibres (22% tumours)	Pott et al. (in press)
Rat	125 (63 males, 62 females)	Intrapleural 3-fold injection of 20 mg artificial Na, Mg-hydroxyamphibole (90% < 5 µm in length)	Mesotheliomas in 54.5%	Pylev et al. (1975)

WL = Working levels.
UICC = Union Internationale Contre le Cancer.
SPF = Specific pathogen free.
JM = Johns Manville.

The carcinogenic potency of glass fibres (JM 104) following intraperitoneal administration was reduced when the fibres were pre-treated with hydrochloric acid (Pott et al., 1984); the authors suggested that this was a function of the lower durability of the acid-treated fibres (Bellmann et al., in press). In a study by Bignon et al. (1983), the proportion of lung cancer was greater in rats exposed by inhalation to radon and then administered intrapleural injections of one of 10 mineral dusts (including glass fibre) than in rats exposed only to radon.

Pigott & Ishmael (1981) observed only a "mild chronic inflammatory response" 12 months following intraperitoneal injections of 20 mg of two commercial grades of refractory alumina fibre. Styles & Wilson (1976) observed nodular deposits of connective tissue, but no fibrosis, following intraperitoneal injection of alumina and zirconia fibre (0.2 ml 10% suspension). On the basis of the results of a study in which 20 mg of various fibrous dusts were injected intrapleurally in rats, Wagner et al. (1973) estimated the carcinogenicity factor (x 10^9) for "ceramic fibre" to be 0.16 compared with 2.28 - 2.35 for SFA chrysotile. Following intraperitoneal administration of 25 mg of fibrous ceramic aluminium silicate glass to rats, Davis et al. (1984) observed tumours in 9.3% of the exposed animals compared with 90% in similar studies with chrysotile. Moreover, time to first tumour was 850 days for ceramic fibres compared with 200 days for chrysotile. However, in a recent study, abdominal mesotheliomas were observed in 83% (19 out of 23) of Osborne-Mendel rats receiving a single intraperitoneal injection of 25 mg of 1.8-μm (mean diameter) refractory ceramic fibres (Smith et al., in press).

The need for caution in the extrapolation of the results of studies involving injection or implantation in body cavities to predict the potency of various fibre samples, even with respect to the induction of mesotheliomas cannot be overemphasized. The relevance of these types of studies to other types of cancer, such as lung cancer, has not been established.

7.2 *In Vitro* Studies

Several important factors that influence the pathogenicity of fibrous dusts *in vivo* (e.g., deposition, clearance, and immunological function) are absent in *in vitro* systems. Moreover, the results of such assays vary considerably, depending on the test system used. However, on the basis of available results for all fibres, it appears that a combination of *in vitro* tests may be useful in predicting the fibrogenicity of fibres *in vivo*. The predictive value of *in vitro* tests for the carcinogenicity of fibrous dusts is, at present, less well established, but there has been some consistency between the results of specific *in vitro* assays and the induction of mesotheliomas in *in vivo* studies involving intrapleural administration. In general, therefore, *in vitro* assays are considered to be useful for the investigation of mechanisms of disease induced by fibre dusts and possibly as preliminary screening tests for pathogenic fibres.

The results of *in vitro* studies on various types of MMMF are presented in Table 20. To date, the effects of glass fibres of various size distributions have been examined in a wide range of systems including bacteria, cultured erythrocytes, fibroblasts and macrophages from several animal species, macrophage-like cell lines and cultured human fibroblasts, erythrocytes, lymphocytes, and bronchial epithelial cells. In most of the assays, cytotoxicity or cytogenetic effects have been a function of fibre size distribution, with longer (generally $> 10\ \mu m$), thinner (generally $< 1\ \mu m$) fibres being the most toxic (Brown et al., 1979b; Lipkin, 1980; Tilkes & Beck, 1980, 1983a,b; Hesterberg & Barrett, 1984; Forget et al., 1986; Hesterberg et al., 1986). In general, "coarse" fibrous glass (with relatively large fibre diameters, e.g., JM 110) has been less cytotoxic in most assays than chrysotile or crocidolite (Richards & Jacoby, 1976; Haugen et al., 1982; Nadeau et al., 1983; Pickrell et al., 1983). However, the cytotoxicity or transforming potential of "fine" glass (e.g., JM 100) has approached that of the asbestos varieties (Pickrell et al., 1983; Hesterberg & Barrett, 1984).

Results concerning the effects of fibre coating on toxicity in *in vitro* assays have been contradictory. Brown et al. (1979a) reported that "clean" samples of rock, glass, and slag wool (i.e., resin removed by pyrolysis) were slightly more cytotoxic for V79-4 and A549 cells than were resin-coated samples of the same materials. In contrast, Davies (1980) found that removal of the binder from rock and slag wools and resin from glass wool had no effect on their cytotoxicity for mouse peritoneal macrophages.

With respect to genotoxicity, glass fibres (JM 100 and 110) did not induce point mutations in *E. coli* or *S. typhimurium* (Chamberlain & Tarmy, 1977). Casey (1983) did not observe any effects of fine (JM 100) or coarse (JM 110) fibrous glass on sister chromatid exchange in CHO-K_1 cells, human fibroblasts, or lymphoblastoid cells. However, a variety of fibrous materials delayed mitosis in human fibroblasts and CHO-K_1 cells. The order of potency in CHO-K_1 cells was chrysotile > fine glass (JM 100) > crocidolite > coarse glass (JM 110). JM 100 glass fibres induced chromosomal breaks, rearrangements, and polyploidy in CHO-K_1 cells, while JM 110 glass fibres did not have any effect (Sincock et al., 1982). Oshimura et al. (1984) reported that glass microfibres (JM 100) induced cell transformation and cytogenetic abnormalities in Syrian hamster embryo (SHE) cells. The cytotoxicity, transforming frequency, and micronucleus induction frequency of glass microfibres (JM 100) in SHE cells

Table 20. *In vitro* studies

Fibre type	Fibre size distribution	Results	Reference
Fibrogenic effects			
Glass fibres	JM 100: mean length, 2.7 µm; mean diameter, 0.12 µm; JM 110: mean length, 26 µm; mean diameter, 1.9 µm	Increase in total protein at 48 h and in total protein and collagen synthesis at 96 h in rat fibroblasts; short fibres more fibrogenic than long ones at 96 h	Aalto & Heppleston (1984)
Glass fibre dust; chrysotile asbestos	Not reported	Initial slight fibrogenic response with glass fibre compared with pronounced effect of chrysotile asbestos on rabbit lung fibroblasts	Richards & Morris (1973)
Alumina and zirconia fibres	Saffil alumina: median length, 17 µm; median diameter, 3.6 µm; Saffil zirconia: median length, 11 µm; median diameter, 2.5 µm	Low cytotoxicity for rat peritoneal macrophages; non-fibrogenic in intraperitoneal injection studies	Styles & Wilson (1976)
Cytotoxicity			
Glass fibres	% of fibres < 0.6 µm in diameter and > 5 µm in length: JM 100T, 36%; JM 110T, 27.4%; JM 100R, 36.7%; JM 110R, 34.6%	Fibres < 10 µm in length not cytotoxic for V79-4 cells, A549 cells, or mouse peritoneal macrophages	Brown et al. (1979b)
Glass fibres	Homogenized GF/D Whatman filters washed in HCl; fibre size distribution not reported	Mildly cytotoxic for cultured human bronchial epithelial cells; chrysotile 1000 times more cytotoxic than glass fibre	Haugen et al. (1982)
Glass fibres	Not reported	Not toxic for human peripheral blood lymphocytes; "large" glass fibre as cytotoxic for mouse peritoneal macrophages as chrysotile; "small" glass fibre not cytotoxic in this assay	Nakatani (1983)

Table 20 (contd).

Fibre type	Fibre size distribution	Results	Reference
Cytotoxicity (contd)			
Glass fibres	"Respirable"; not reported	Morphological changes and alterations in reticular deposition in rabbit lung fibroblasts less severe for chrysotile	Richards & Jacoby (1976)
Glass fibres	JM 100: mean diameter, 0.23 µm (99.8% < 10 µm long); JM 104: mean diameter, 0.32 µm (99.1% < 10 µm long); GFF: mean diameter, 0.43 µm (99.1% < 20 µm long); GfF: mean diameter, 0.19 µm (99% < 3 µm long)	Toxicity in phagocytic ascites tumour cells from Wistar rats correlated with fibre lengths and diameter	Tilkes & Beck (1980)
Glass fibres	1. 83.8% < 1 µm long; 100% < 0.3 µm wide 2. 75% > 5 µm long; 99.7% < 0.3 µm wide 3. 15.5% > 5 µm long; 22.3% < 0.3 µm wide 4. 99.3% > 5 µm long; 82.9% < 0.3 µm wide 5. 44.1% > 5 µm long; 3.4% < 0.6 µm wide 6. 90.1% > 5 µm long; 4.1% < 0.6 µm wide	Long narrow glass fibres as toxic for phagocytic ascites tumour cells from Wistar rats as chrysotile; fibres with diameters 3 µm, non-toxic; toxicity increased with increasing fibre length	Tilkes & Beck (1983a)
Glass fibres	Owens Corning beta-fibre: 99% > 10 µm long; 3% < 3 µm wide; JM 100: 46% > 10 µm long; 50% < 0.2 µm wide	Significant increase in squamous cell metaplasia and labelling index in organ cultures of hamster trachea (similar to crocidolite)	Woodworth et al. (1983)

Table 20 (contd).

Glass fibres, glass wool	Not reported; glass fibres from "GF/C microfilter"	Some binding of carcinogens (benzo(a)-pyrene, nitrosonornicotine, N-acetyl-2-aminofluorene) by fibrous glass and glass wool, but less than that for most other mineral fibres examined, such as chrysotile and attapulgite; negligible haemolysis in sheep erythrocytes or cytotoxicity in P388 D_1 cells	Harvey et al. (1984)
Three types of glass fibre-containing household insulation; microfibre insulation material	Microfibre insulation: CMD, 0.1 - 0.2 µm; household insulation, CMD, 2 - 4 µm	Cytotoxicity of microfibre insulation in pulmonary alveolar macrophages from beagle dogs similar to that of crocidolite; household insulation fibres non-toxic	Pickrell et al. (1983)
Glass fibres; rock, slag, and glass wool	JM 100, JM 110, and JM 100 (respirable); size distribution not reported	"Fine" glass fibres (JM 100) more cytotoxic for mouse peritoneal macrophages than UICC crocidolite, but "coarse" glass fibre had little activity; removal of binder from rock and slag wools and resin from glass wool had no effect	Davies (1980)
Glass fibres	JM 104; 90% < 15 µm in length and 0.5 µm in diameter	Induced a small release of beta-galactosidase, but was not cytotoxic for rabbit alveolar macrophages	Jaurand et al. (1980)
Glass fibres	Not reported	"Long" fibres produced a protracted pathogenic but not cytotoxic effect in guinea-pig alveolar macrophages; for short fibres, morphological pattern similar to that for inert dusts	Bruch (1974)
Glass fibres	JM 100-milled to alter fibre lengths (size distribution not reported)	Glass fibres induced dose-dependent release of prostaglandins and beta-d-glucuronidase from perfused rat alveolar macrophages, cell aggregation and mortality; long fibres more active than short ones	Forget et al. (1986)

Table 20 (contd).

Fibre type	Fibre size distribution	Results	Reference
Cytotoxicity (contd)			
Glass fibres	"Pyrex glass fibres"; size distribution not reported	In contrast to chrysotile, glass fibres did not promote fusion of human erythrocyte membranes or haemolysis and fusion of fowl erythrocytes	Ottolenghi et al. (1983)
Glass fibres	Fibre size distribution as for Tilkes & Beck (1983a)	Cytotoxicity in guinea-pig- and rat-lung macrophages related mainly to fibre size distribution; depression of phagocytosis	Tilkes & Beck (1983b)
Glass fibres, glass wool	Not reported	Glass fibre (GF/C filters) less cytotoxic for 3T3 fibroblasts than several forms of chrysotile and attapulgite but more cytotoxic than UICC crocidolite; minimal cytotoxicity of Pyrex (glass wool)	Dumas & Page (1986)
Glass fibres	JM 100 - mean length, 15 µm; mean diameter, 0.2 µm; JM 100 (milled) - mean length, 2 µm; mean diameter, 0.2 µm	Cytotoxic in primary cultures and permanent cell line of rat tracheal epithelial cells; toxicity reduced with milling of the fibres	Ririe et al. (1985)
Glass fibres, glass powder	Glass fibres: lengths ranged from 1 to 20 µm and diameters from 0.25 to 1 µm; glass powder particle size < 3 µm	Glass fibres increased cell permeability of guinea-pig alveolar and peritoneal macrophages; glass powder had little effect	Beck et al. (1972); Beck & Bruch (1974); Beck (1976a,b)
Glass fibres, glass particles	JM 100 glass fibres and particles; size distribution not reported	Glass fibres induced ornithine decarboxylase activity (ODC) in hamster tracheal epithelial cell cultures, while glass particles did not	Marsh et al. (1985)
Borosilicate glass fibres	Fibre size distribution as for sample used by Stanton & Layard (1978)	Cytotoxicity in P388 D_1 macrophage-like cells correlated well with potency in tumour induction reported by Stanton & Layard (1978) in intrapleural implantation studies	Lipkin (1980)

Table 20 (contd).

Cytotoxicity (contd)			
Rock wool, slag wool, and glass wool	Not reported	Cytotoxic for V79-4 and A549 cells; "clean" (resin removed by pyrolysis) samples slightly more active than resin-coated variety	Brown et al. (1979a)
"Ceramic fibre"	Elutriated sample; 89% of fibre lengths < 5 µm; fibre diameters < 3 µm; 7.4 fibres/10^{-10} g (small number compared with different types of chrysotile)	Ceramic fibre less cytotoxic for P388 D_1 cells than chrysotile samples and most of the amphibole samples	Wright et al. (1986)
Ceramic	Not reported	Ceramic fibres were inert in the V79/4 cell colony inhibition assay, but increased the diameter of the A549 cells	Brown et al. (1986)
Xonotlite (synthetic)	Diameter, ≤ 0.1 µm; length, < 2 µm (100%)	Fibres phagocytosed by primary rat hepatocytes in culture, based on ultrastructural analysis	Denizeau et al. (1985)
Mineral wool, glass fibre	Mineral wool: D_f = 3.3 µm, L_f = 221.4 µm; glass fibre: D_f = 0.2 µm, L_f = 11.4 µm	Mineral wool not cytotoxic for rat alveolar macrophages, nor haemolytic for rat erythrocytes, but increased oxidant production in rat macrophages; glass microfibres induced haemolysis, but to a lesser extent than chrysotile	Nadeau et al. (1983)

Table 20 (contd).

Fibre type	Fibre size distribution	Results	Reference
Genetic and related effects			
Glass fibres	JM 100 (fine) and JM 110 (coarse); size distribution not reported	No effects on sister chromatid exchange in CHO-K_1 cells, human fibroblasts, or lymphoblastoid cells; mitotic delay in CHO-K_1 cells and human fibroblasts; order of induced delay in CHO-K_1 cells: chrysotile > fine glass > crocidolite > coarse glass	Casey (1983)
Glass fibres	JM 100: mean length, 9.5 µm; mean diameter, 0.13 µm; shorter fibre samples of similar diameter produced by milling; JM 100: mean diameter, 0.8 µm	JM 100 glass fibres as active as chrysotile in transforming Syrian hamster embryo cells; thick glass fibres (JM 110) 20 times less potent than thin (JM 110) ones; 10-fold decrease in transformation with reduction of fibre length from 9.5 to 1.7 µm (JM 100); no transformation for fibre length of 0.95 µm	Hesterberg & Barrett (1984)
Glass fibres	"AAA" glass fibre of different size distributions	Sizes of glass fibres that promote growth in BHK 21 or 3T3 fibroblasts (> 20 µm in length) similar to sizes that induce mesotheliomas following intrapleural implantation	Maroudas et al. (1973)
Glass fibres	JM 100-milled (average length, 2.2 µm) and unmilled (average length, 15.5 µm) (95 - 98% of fibre diameters < 0.5 µm)	Glass fibres phagocytosed by Syrian hamster embryo cells and accumulated in perinuclear region of the cytoplasm; milling of glass fibres (resulting in a 7-fold decrease in length) reduced percentage of phagocytosis, cytotoxicity, transformation frequency, and micronucleus induction frequency	Hesterberg et al. (1986)

Table 20 (contd).

Glass fibres	JM 100 (mean length, 2.7 µm, mean diameter, 0.12 µm); JM 110 (mean length, 26 µm; mean diameter, 1.9 µm)	No mutagenic activity in bacterial strains of *Escherichia coli* or *Salmonella typhimurium* at levels of up to 1000 µg/plate	Chamberlain & Tarmy (1977)
Glass fibres	JM 100 (mean diameter, 0.1 - 0.2 µm; mean length varied by milling); JM 110 (mean diameter, 0.8 µm)	JM 100 fibres induced cell transformation and cytogenetic abnormalities in Syrian hamster embryo cells; JM 110 fibres and milled JM 100 fibres much less potent for both end-points	Oshimura et al. (1984)
Glass fibres, glass powder	1.5 - 2.5 µm diameter; fibre size distribution described by Wagner et al. (1973)	Exposure of CHO-K_1 cells to one concentration of glass fibre did not increase the frequency of chromosomal aberrations or polyploid cells	Sincock & Seabright (1975)
JM 100 glass fibres, JM 110 glass fibres	Mean particle lengths between 2.7 and 26 µm; mean particle diameters between 0.12 and 1.9 µm	JM 100 fibres caused chromosomal breaks, rearrangements, and polyploidy in CHO-K_1 cells; no effects with JM 110 fibres	Sincock et al. (1982)
JM glass fibres	Not reported	Dose-dependent interference with development of limb buds in 11-day-old mouse embryos in culture	Krowke et al. (1985)

JM - Johns Manville.
CMD - Count median diameter.
CHO - Chinese hamster ovary.

was reduced by milling (i.e., decreasing the length) (Hesterberg et al., 1986).

Few data are available on the toxicity of MMMF other than glass fibres in *in vitro* assays. However, Styles & Wilson (1976) reported that the cytotoxicity of Saffil alumina and Saffil zirconia fibres (size range unspecified) in rat peritoneal macrophages was low; this agreed well with the lack of fibrogenicity observed in intraperitoneal injection studies with the same materials. Wright et al. (1986) reported that "ceramic fibre" (type and source unspecified) was less cytotoxic for P_{388} D_1 cells than chrysotile and the amphiboles.

7.3 Mechanisms of Toxicity - Mode of Action

The results of toxicological studies have demonstrated the importance of fibre dimension, persistence, and durability in the pathogenesis of fibrous dust-related diseases. Surface charge and chemical composition may also play an important role. However, the mechanisms by which respirable fibrous materials cause fibrosis and cancer are not well understood. The sequence of cellular events leading to fibrosis has been hypothesized on the basis of observations in animals after inhalation or intratracheal instillation of asbestos (Davis, 1981). It is likely that the sequence of events in the induction of fibrosis by other fibres of similar dimensions and durability is similar. Short fibres deposited on the alveolar surface are phagocytosed by macrophages and removed by the mucociliary escalator. Fibres longer than 10 μm are often surrounded by groups of macrophages, which may fuse to form multinucleated giant cells. Some of the dust-containing macrophages become incorporated into the lung parenchyma and die, and the fibres are rephagocytosed by new populations of macrophages. The presence of these fibres in distal airways stimulates excess deposition of collagen and reticulin fibres.

The deposition of fibres in the pulmonary region of the respiratory tract is thought to be important only if the fibres can penetrate into the interstitium where interstitial macrophages can phagocytose them. In the process, the macrophage is impaired; this may "trigger" adjacent fibroblasts in the interstitium to start producing more collagen, leading eventually to fibrosis.

Several mechanisms by which asbestos and possibly other fibrous materials cause cancer have been suggested. Because of the lack of activity of most of these materials in gene mutation assays, it has been suggested that they may act by epigenetic mechanisms (NRC, 1984). For example, it has been proposed that asbestos acts primarily as a promoter or cocarcinogen or that cancer occurs secondarily to the induction of inflammation or fibrosis. However, results of *in vitro* studies in which chromosomal aberrations have been induced by glass fibres (Oshimura et al., 1984), suggest that the fibres can directly affect the genetic material.

Generation of reactive oxygen metabolites by fibrous materials in *in vitro* assays has also been observed and has been postulated to play a role in cytotoxicity (Goodglick & Kane, 1986). It may also be that fibrogenesis and carcinogenesis may be a result of several of these mechanisms.

8. EFFECTS ON MAN

8.1 Occupationally Exposed Populations

8.1.1 Non-malignant dermal and ocular effects

The effects of MMMF on the skin were recognized as early as the turn of the century (HMSO, 1899, 1911). Fibrous glass and rock wool fibres (mainly those greater than 4.5 - 5 μm in diameter) cause mechanical irritation of the skin characterized by a fine, punctate, itching erythema, which often disappears with continued exposure (Hill, 1978; Björnberg, 1985). Possick et al. (1970) described the effects of short glass fibres temporarily piercing the epidermis producing sensations varying from itching and burning to pain. They suggested that skin penetration was directly proportional to fibre diameter and inversely proportional to fibre length. The primary lesion is a papule or papulo-vesicle, and the authors quote histological reports of oedema of the upper dermis with round cell infiltrates. Secondary lesions include bacterial infections, which develop as a result of scratching, and lichenification. Urticaria occurs in dermatographic subjects.

Data concerning the incidence or prevalence of dermatitis in workers exposed to MMMF are sparse. Sixty-one percent of 62 glass wool workers in a Swedish plant had cutaneous signs or symptoms at the end of a working day; 45% had visible dermal lesions (Björnberg, 1985). Twenty-five percent of 315 subjects exhibited skin reactions at 72 h when patch tested for 48 h with rock wool (Björnberg & Löwhagen, 1977) and, in a study on 135 Italian fibrous glass production workers, the prevalence of dermatitis was reported to be 19% (Arbosti et al., 1980). However, according to Gross & Braun (1984), much lower prevalences were reported in investigations previously conducted in the USA. In a study on 467 glass wool workers, Maggioni et al. (1980) reported that 14% had skin disease, mainly primary irritative dermatitis.

There are also few reliable data concerning the tolerance of workers to MMMF-induced dermatitis. Björnberg et al. (1979) described a group of 60 workers at a glass wool factory; after counselling, 38 were able to continue work but 22 had to be transferred to work with less potential for fibre exposure; 6 of these workers eventually had to leave their jobs because of this irritation.

The irritant dermatitis induced by MMMF may be complicated by an urticarial and eczematous reaction that sometimes mimics an allergic response, both clinically and histologically. In addition, allergic reactions to resins used in MMMF production occasionally occur (Fisher, 1982). For example, allergic dermatitis in 54% of 160 workers engaged in glass fibre manufacture was attributed to exposure to epoxy resin (Cuypers et al., 1975); however, coating with phenol-formaldehyde did not have any effects on skin reactions induced by rock wool (Björnberg & Löwhagen, 1977). Conde-Salazar et al. (1985), in their case report of severe dermatitis in a glass fibre spinner sensitized to epoxy oligomer, commented that glass fibres *per se* produced only an irritant dermatitis.

Until recently, reports of eye irritation in populations exposed occupationally to MMMF were restricted to a few isolated cases in the early literature (Gross & Braun, 1984). However, in a recent Danish study, the frequency of eye symptoms significantly increased and the number of microepithelial defects on the medial bulbar conjunctiva and, in some cases, the neutrophil count of the conjunctival fluid were increased after 4 days of exposure in 15 rock wool workers (employed for > 6 months) compared with controls matched for age, sex, and smoking habits. While there were no ophthalmological differences between exposed workers and controls on Monday morning, an excess of mucous was found in exposed workers, suggesting that the effect was not completely reversible over the weekend (Stokholm et al., 1982).

8.1.2 Non-malignant respiratory disease

In reports that appeared in the early literature, several cases of acute irritation of the upper respiratory tract and more serious pulmonary diseases, such as bronchiecstasis, pneumonia, chronic bronchitis, and asthma were attributed to occupational exposure to various MMMF. On the basis of recent reviews of these isolated reports, several authors (Hill, 1978; Upton & Fink, 1979; Saracci, 1980; Gross & Braun, 1984; Wright, 1984) have concluded that exposure to MMMF was probably incidental rather than causal in most of these cases, since the reported conditions have not been observed in most of the more recently conducted epidemiological studies. These case reports are not considered further here.

The epidemiological studies of non-malignant respiratory disease in populations exposed occupationally to MMMF are mainly of two types: cross-sectional investigations of the prevalence of signs and symptoms of respiratory disease and historical prospective (cohort) studies of mortality due to non-malignant respiratory disease.

8.1.2.1 Cross-sectional studies

The design and results of the available cross-sectional studies of respiratory effects in MMMF production workers are summarized in Table 21. Most of these studies have involved one

Table 21. Cross-sectional studies of respiratory effects in MMMF-exposed workers[a]

Study protocol	Results	Comments	Reference
Examination of chest radiographs of 1389 employees in a glass wool manufacturing plant (> 10 years exposure to dust concentrations from 0.93 to 13.3 mg/m^3)	No unusual pattern of radiological densities; frequency no higher in those with greatest exposure than in those with the least	*Ad hoc* radiological assessments by one reader difficult to interpret epidemiologically	Wright (1968)
Examination of chest radiographs of 2028 male workers in the fibrous glass plant studied by Wright (1968) (two-thirds employed for > 10 years)	Radiographic abnormalities in 16% of workers; strength of correlation between prevalence and mean duration of employment and between prevalence and mean age similar; no difference in prevalence between office and production workers	Not possible to interpret relationship between response and duration of employment, because of confounding with age in grouped data; within-group analyses or logistic regression analyses of data could be useful	Nasr et al. (1971)
Examination of respiratory symptoms (determined by questionnaire), radiographic changes, and pulmonary function (VC, FEV_1, and VC/FEV_1) in 232 employees in the fibrous glass plant studied by Wright (1968) and Nasr et al. (1971); for sub-sample of 30, comprehensive physical examination, chest fluoroscopy, electrocardiogram, VC, FVC_1, FVC_2, FVC_3, mid-expiratory flow rate, maximal voluntary ventilation, residual volume, and pulmonary diffusing capacity	No significant difference in the prevalence of chronic respiratory illness between highest and lowest exposure category	Numerical data to support conclusions not provided	Utidjian & Cooper (1976)

Table 21 (contd).

Comparison of respiratory symptoms (determined by MRC questionnaire), radiographic changes, and pulmonary function (peak expiratory flow, FEV_1, and FVC) in 70 fibrous glass wool workers (exposed for a mean period of 19.85 years; mean level, 0.9 respirable fibres/cm^3) and controls matched for age, sex, height, and weight, and living in the same geographical area	No evidence of pulmonary effects due to exposure to fibrous glass, but 45% of exposed population had suffered for a short period from "new starter's glass-fibre rash"	Well-controlled study of a "survivor population" of current employees	Hill et al. (1973)
Examination of respiratory symptoms (determined by MRC questionnaire), radiographic changes, and pulmonary function (VC, FEV) in 340 workers (> 10 years exposure) from a glass wool manufacturing plant	Some indication of chronic non-specific lung disease related to level but not to length of exposure to fibres; prevalence of small opacities related to environmental pollution (never having lived in a smoke-free zone), and to previous employment in dusty occupations, but no significant relationship with either length or level of exposure to MMMF	Only study that includes ex-employees; absence of relationship with indices of duration of exposure suggests possible confounding with exposure to other dusts	Hill et al. (1984)

Table 21 (contd).

Study protocol	Results	Comments	Reference
Examination of respiratory symptoms (determined by ATS questionnaire), radiographic changes, and pulmonary function (FVC, FEV_1, FEV_1/FVC, FEF 25-75, RV, VC, TLC, DL, DL/ALVOL) in 1028 workers in 7 glass wool and mineral wool plants exposed for a median period of 18 years to median levels from < 0.032 to 0.928 fibres/cm^3	No relationship between respiratory symptoms or adverse effects on pulmonary function and exposure; greater prevalence of small opacities among current smokers in 2 plants producing ordinary and fine fibres; for non-smokers, greater prevalence observed in only 1 of these plants; among smokers, prevalence of small opacities increased with increasing length of employment, and authors concluded that exposure to MMMF with small diameters may lead to low-level profusion of opacities	Results suggest that fibres with small diameters have an effect on the lung parenchyma; sound, well-documented study; replication and confirmation desirable	Weill et al. (1983, 1984)
Survey of workers with > 1 year employment at 2 glass wool factories (A and B); age- and smoking habit-stratified sample from factory A (n = 367); all eligible from factory B (n = 157); respiratory symptoms (questionnaire), clinical examinations, chest radiography, spirometry, CO-transfer factor for those working with fibres compared with those not exposed at same factory (factory A: 114, factory B: 34)	Prevalence of breathlessness and nasal cavity symptoms significantly higher in exposed workers at factory A only; no other significant differences; bronchitic symptoms a little more frequent in both exposed groups ($P > 0.05$); in nested case-control study, a significantly shorter mean period of employment for cases in factory A	No convincing evidence that exposure to fibres affects respiratory system	Moulin et al. (1987)

Table 21 (contd).

Examination of exposure and respiratory symptoms (determined by self-administered questionnaire) in 135 000 male Swedish construction workers; separate analysis for "former", "never", and present smokers; control for age and asbestos exposure	Positive association between handling of MMMF and bronchitis	Evidence of errors in questionnaire-determined exposure to asbestos undermines interpretation of the association as causal	Engholm & Von Schmalensee (1982)
Examination of chest radiograms of 84 workers in a rock and slag wool plant (exposure periods, 7 - 26 years)	No radiographic evidence attributable to mineral wool exposure	Radiological procedures not documented adequately	Carpenter & Spolyar (1945)
Examination of respiratory symptoms (determined by self-administered questionnaire) in 537 insulators working in Quebec in 1982 without asbestosis; logistic regression analysis with smoking, reactivity before employment and occupational exposure to dust and mineral fibre as risk factors (workers without asbestosis in 558 respondents out of 644 identified subjects)	Wheezing and breathlessness related primarily to current smoking and asthmatic predisposition antedating work; these symptoms also related to occupational exposure in subjects with prior airways hyperreactivity	Effects of work-place exposure may have been underestimated due to selective withdrawal from the active work-force and to inaccuracies in exposure (based on duration of employment only); not possible to separate exposure to asbestos, MMMF, and dust	Ernst et al. (1987)
Examination of respiratory symptoms (determined by questionnaire) and spirometry in 235 workers in a continuous filament glass fibre plant (83% of the shop floor workers and 21 others who were infrequently or never on the shop floor); prick testing and bronchial challenge in some workers identified as having work-related asthma	7 cases of "work-related asthma" identified	No control data; cause of the asthma not identified	Finnegan et al. (1985)

8

Table 21 (contd).

Study protocol	Results	Comments	Reference
Examination of respiratory symptoms (determined by ATS questionnaire) and pulmonary function (FVC, FEV_1) in > 4000 workers in 14 US and Canadian fibrous glass (type unspecified) plants exposed for a median period of 7.7 years (males) to < 0.5 fibres/cm^3	No relationship between chronic cough, phlegm, dyspnoea, or effects on pulmonary function and length of employment; respiratory symptoms related to smoking and pulmonary function related to smoking and age	Preliminary results of a longitudinal investigation	Lockey (in press)
Comparison of respiratory symptoms (determined by questionnaire), radiographic changes, and pulmonary function in 467 glass wool workers (exposed for a mean period of 13 years) and in controls	Higher incidence of chronic and dysplastic pharyngolaryngitis in workers (> 5 years) exposed to highest concentrations; skin disease (mainly primary irritative dermatitis) in 14% of the working population		Maggioni et al. (1980); Saracci & Simonato (1982)
Comparison of respiratory symptoms (determined by questionnaire) and pulmonary function (lung volume, airway resistance, forced expiratory flow, closing volume, transfer factor) in 21 employees (> 45 years old) of a rock wool manufacturing plant (mean exposure for 17.6 years to 0.19 respirable fibres/cm^3) and in 43 controls (50 - 70 years of age) (values corrected for differences in age, body weight, and smoking habits)	No significant differences in pulmonary function between exposed and control groups; some symptoms of chronic bronchitis in exposed group	Not possible to generalize from careful, detailed physiology in only 21 subjects	Malmberg et al. (1984)

Table 21 (contd).

Spirometry and determination of lung volumes, closing volumes, slope of the alveolar plateau, maximum expiratory flow in air and after helium-oxygen breathing and elastic recoil pressures in 8 sheet metal workers exposed to fibrous glass compared with 7 sheet metal workers "never exposed" to fibrous glass (workers with no history of smoking, pleural plaques, or asbestos exposure chosen from 251 respondents out of 532 workers surveyed)	Slightly increased elastic recoil in exposed subjects; authors could not exclude the possibility that fibrous glass could cause a faint and probably harmless fibrous reaction in the lung parenchymma	Length or extent of exposure unknown; bias may have been introduced by low (47%) response rate, and hyper-critical inclusion criteria	Sixt et al. (1983)
Examination of pulmonary function (FVC, FEV_1, and MEF 50%) in 162 rock wool workers (exposed to 0.003 - 0.463 respirable fibres/cm^3); 5 years later, examination of pulmonary function in 102 people from original population exposed only to MMMF in the interval	In original study group, mean pre-shift values of FVC, FEV_1, and MEF 50% significantly lower than reference values, but no relationship with cumulative exposure; in second study, no change in pulmonary function associated with MMMF exposure	Longer follow-up required to detect possible exposure-related lung function decrements	Stahuljak-Beritic et al. (1982)

[a] Abbreviations
MRC = Medical Research Council (United Kingdom).
ATS = American Thoracic Society.
FEV_1 = Forced expiratory volume in 1 second.
FVC = Forced vital capacity.
FEF 25-75 = Forced expiratory volume during the mid-half of the forced vital capacity.
RV = Residual volume.
VC = Vital capacity.
TLC = Total lung capacity.
DL = Diffusing capacity.
DL/ALVOL = Ratio of diffusing capacity to alveolar volume.
MEF 50% = Maximal expiratory flow rate at 50% of forced vital capacity.

or a combination of the following elements: determination of respiratory symptoms through administration of questionnaires, pulmonary function testing, and radiological examination of the lung. Limitations that should be borne in mind in reviewing the results of cross-sectional studies include the following: persons who have left employment because of ill health are selectively excluded and health effects are monitored at only one particular time (i.e., prevalence rather than incidence is determined). Nevertheless, such investigations do provide useful preliminary data.

Hill et al. (1973) compared radiographic appearances, reports of respiratory symptoms, and lung function measurements in 70 workers who had been exposed at a glass wool factory in England for an average of nearly nearly 20 years, with those of other workers, matched for sex, age, height, and weight, who were from the same geographical area but had not been exposed to fibrous dust. The frequency of any abnormalities among those exposed was no higher than in the controls. Hill et al. (1984) later examined 74% of 340 workers and ex-employees, in the 55 - 74 year age group, who had been employed for more than 10 years at the same factory, and also samples of other groups of employees and ex-employees for whom participation rates were low (121/250 overall). The results of the second study indicated some occupationally related chronic non-specific lung disease in the 340 fibrous glass workers examined. The statistically significant impairment of lung function was reported to be more severe among workers in occupational groups designated as likely to have had high exposure to fibres, but there was no relationship with length of time employed in these groups. Eleven percent of all 340 workers studied had small opacities on their chest radiographs of profusion category higher than 0/1 in the ILO (1980) international classification. A subgroup of 161 men, who had no history of occupational exposure to any dusts other than glass fibres, included 15 (9%) with category 1/0 or higher, but there was no statistically significant relationship between the occurrence of these signs and either intensity or duration of exposure to glass fibres. The factory concerned is included in the European study of mortality discussed in sections 8.1.2.2 and 8.3.

Moulin et al. (1987) conducted a cross-sectional survey of workers in two glass wool manufacturing plants in France. A total of 524 subjects (367 in factory A and 157 in factory B) were examined for respiratory symptoms, radiological pulmonary changes, and functional respiratory alterations. In each factory, the prevalences of abnormalities among those exposed to fibres were compared with findings in unexposed workers (administrative, oven, and polystyrene workers). The only statistically significant differences between the two groups were higher prevalence rates of breathlessness and nasal cavity symptoms in the exposed workers at factory A. The prevalence of chronic bronchitis was increased slightly among exposed workers at both factories. No other exposure-related effects on the respiratory system were observed. The authors also used a case-control approach in which cases were the subjects in the quartile showing the worst results from the various examinations and the age-matched controls were chosen from among the others. There was no difference as to duration of exposure in factory B, while cases had a shorter period of employment in factory A. The results of this study do not indicate any consistent effects of exposure to glass wool fibres on the respiratory system.

Weill et al. (1983, 1984) reported results based on a very detailed statistical analysis of observations on 1028 men working at 7 North American glass wool or mineral wool factories. All 7 factories were included in the comprehensive US mortality study discussed in sections 8.1.2.2 and 8.1.3. A statistically significant increase in the chance of finding small opacities (category 0/1 or higher) with increasing length of employment at the factories was found among some 450 current cigarette smokers. The highest profusion category on the ILO (1980) scale, as determined from the median of independent assessments by 3 readers on 6 films, was category 1/1. There was no evidence in this study that exposure to fibres increased the prevalence of respiratory symptoms or impairment of lung function.

Three reports refer to separate cross-sectional surveys, carried out at different times, of workers at one large factory producing continuous filament glass fibres (Wright, 1968; Nasr et al., 1971; Utidjian & Cooper, 1976). The last survey was of a stratified approximately 10% random sample of the workers and included the use of a respiratory symptom questionnaire and measurement of lung function. The earliest of these studies (Wright, 1968) did not include those who had been employed by the company for less than 10 years, but Nasr et al. (1971) studied nearly all male employees (N = 2028), including 196 office workers. Radiographic appearances were characterized using different conventions.

Wright (1968) did not find any unusual radiological patterns, and the frequency of various radiological appearances among those characterized subjectively as "likely to have had highest exposure" was no greater than in those considered to have least exposure. Nasr et al. (1971) reported very similar prevalences of radiographic abnormalities among the office and production workers (17% and 16%, respectively). Mean ages and mean durations of employment in 6 age groups were correlated almost perfectly, and these mean values were each correlated positively with the prevalence rates in the age groups. Thus, there was almost total confounding between the average age and duration of employment in the grouped data.

Utidjian & Cooper (1976) failed to demonstrate any significant differences in the prevalence of chronic respiratory illness or impairment of lung function between categories of workers designated as likely to have had the highest and lowest exposures to airborne fibres in the same plant. However, neither tabular nor graphical presentations of results were included in their brief report.

Engholm & Von Schmalensee (1982) collected data, through self-administered questionnaires, on occupational histories, smoking habits, and respiratory symptoms in 135 000 men working in the Swedish construction industry. Reports of respiratory symptoms indicative of chronic bronchitis were grouped according to age, duration of exposure to MMMF, smoking habits, and whether or not there had been any occupational exposure to asbestos. Standardized prevalence rates of bronchitic symptoms, adjusted for age and for histories of exposure to asbestos, increased consistently with increasing years of exposure to MMMF in each smoking category. The authors discuss likely sources of bias in their results, including the possibility that factors not considered in their analysis may have been correlated with duration of exposure to MMMF and with the occurrence of bronchitic symptoms (confounding). They argue that their results are unlikely to be explained in this way. This argument is weakened by their observation of a possible association between bronchitis and exposure to asbestos in their data and subsequent demonstration of substantial errors in the self-administered-questionnaire-determined exposures for asbestos (Engholm et al., in press).

Other studies listed in Table 20, with apparently negative or equivocal results, are even more difficult to interpret epidemiologically, because of: the small numbers of persons studied, the selection criteria adopted to justify inclusion in analyses, the absence of information on likely intensity or duration of exposure to fibres, or incomplete documentation of study procedures and findings.

In summary, results from available cross-sectional studies indicate that occupational exposure to various types of MMMF may be associated with adverse effects on the respiratory tract, but no consistent pattern has emerged. Findings suggesting this association require confirmation, preferably in longitudinal investigations of exposed cohorts, including ex-employees. In particular, follow-up of the observed increased prevalence of small opacities in smokers exposed to MMMF would be valuable.

8.1.2.2 Historical prospective studies

The design and results of relevant historical prospective analytical epidemiological studies are presented in Table 22.

Table 22. Historical prospective (cohort) and case-control studies of MMMF-exposed workers

Study protocol	Results[a]	Comments	Reference
16 661 male workers in 11 fibrous glass (6 glass wool, 3 glass filament, and 2 mixed) and 6 mineral wool plants in the USA employed for 1 year (6 months at 2 plants) between 1945 and 1963 (1940-63 for 1 plant) followed up to 1982 (98% traced; 4986 deaths; caused determined for 97% of deaths); comparison with age- and calendar-adjusted mortality rates for males in the USA or county (malignant neoplasms only); exposure estimate based on environmental survey and information concerning changes in control over the study period	Compared with local rates, no excess of mortality from NMRD for glass wool or continuous filament workers, statistically nonsignificant excess in mineral wool workers; significant excess in all malignant neoplasms and lung cancer 20 years or more after first employment (malignant neoplasms, O:E = 735:678; lung cancer, O:E = 288:255.5) compared with local rates; excess of respiratory cancer greatest in mineral wool workers (O:E = 45:34.4 for those with > 20 years from first exposure); in workers with 30 years from first exposure, excess of respiratory cancer in all but continuous filament production (significant only in mineral wool workers); excess (nonsignificant) of respiratory cancer among 1015 workers "ever exposed" in the production of small diameter fibres compared with those "never exposed" in this sector (O:E = 22:17.8; for those with > 30 years after first exposure, O:E = 6:3.03); little relationship	Asbestos may have been used in some of the mineral wool plants; race unknown for about 30% of the cohort, so white male death rates used for expected values; average exposure for glass filament, 0.011 fibres/cm^3; for glass wool, 0.033 fibres/cm^3; for mixed fibrous glass, 0.059 fibres/cm^3; and for mineral wool, 0.352 fibres/cm^3; 147 of the 301 respiratory cancer deaths 20 years from first exposure occurred at one plant (validity of use of local rates for this plant examined); smoking survey showed smoking habits of MMMF workers similar to those of US white males	Enterline et al. (1983, in press); Enterline & Marsh (1984)

Table 22 (contd).

Study protocol	Results[a]	Comments	Reference
contd.	between respiratory cancer and duration of employment, time from first exposure, or estimated cumulative exposure, but sharp increase in mineral wool workers with date of hire; 3 unconfirmed mesotheliomas in entire cohort (within expected range)		Enterline et al. (1983, in press); Enterline & Marsh (1984)
Case-control study of all MMMF workers from above cohort who died of NMRD or respiratory cancer between January 1950 and December 1982 (cases) and a 4% random sample of workers stratified by plant and year of birth (controls)	Statistically significant relationship between estimated cumulative exposure and respiratory cancer in mineral wool, but not in fibrous glass workers after control for smoking (determined by telephone interviews of workers or their families)		Enterline et al. (in press)
21 967 workers in 13 MMMF plants (7 rock wool, 4 glass wool, and 2 continuous filament) in 7 European countries followed from first employment (1900-55) to 1982 (95% traced; 2719 deaths; causes determined for 98.3% of deaths) (> 1 year employment in English and Swedish plants); comparison with national mortality and, where available, incidence specific for calendar period, sex, and age, and adjusted, in some cases, for local variations; historical environmental investigation	No excess in NMRD compared with national rates (O:E = 165:164.9) in total cohort or any sector; significant excess of mortality from all neoplasms (O:E = 661:597.7), due mainly to significant excess of lung cancer (O:E = 189:151.2) and non-significant excess in cancers of the buccal cavity and pharynx (O:E = 13:10.6), rectum (O:E = 33:27), larynx (O:E = 9:6.3), and bladder (O:E = 24:17.9); no relationship with time since first expo-	Largest cohort studied to date; advantage that some cancer incidence data available; however, workers with < 1 year of employment comprised about 1/3 of the cohort; past manufacture of glass wool in one of the continuous filament plants; in "early technological phase" of rock wool/slag wool production, airborne fibre concentrations higher due to absence of dust suppressing agents; for glass wool, however, changes in airborne fibre concentrations between early and "intermediate technological	Saracci et al. (1984a,b); Simonato et al. (1986a,b, in press)

Table 22 (contd).

	sure for any of the tumour types except lung cancer (O:E = 29:16.8 for > 30 years from first exposure) and bladder cancer; increase in mortality from lung cancer for all sectors (continuous filament, glass wool, and rock wool/slag wool) compared with national rates with relation with time from first exposure for rock wool/slag wool and glass wool subcohorts; when adjusted for local rates, increase in mortality from lung cancer for rock wool/slag wool workers, only, related to time from first exposure; no relationship of lung cancer with duration of employment; excess of lung cancer greatest in rock wool/slag wool workers employed in the "early technological phase" (SMRs = 257 and 214 for comparison with local and national rates, respectively); for bladder cancer, excess in glass wool and rock wool/slag wool workers related to time from first exposure in rock wool/slag wool production; lung cancer incidence increased in rock wool/slag wool subcohort related to time from	phase" less marked due to opposing effects on airborne fibre levels of use of dust suppressants and reduction of fibre diameters; in the "early technological phase" of rock wool/slag wool production, use of arsenic-containing slags, and poor ventilation (resulting in potential exposure to PAHs); data on smoking habits not available; some use of asbestos in some plants	Saracci et al. (1984a,b); Simonato et al. (1986a,b, in press)

Table 22 (contd).

Study protocol	Results	Comments	Reference
contd	first exposure and the early technological phase; significant excess incidence of cancer of the buccal cavity and pharynx in rock wool/slag wool production sector with some (nonsignificant) relationship with time from first exposure; one case of mesothelioma in the cohort (within number expected and worker employed for < 1 year)		Saracci et al. (1984a,b); Simonato et al. (1986a,b, in press)
2557 men with > 90 days employment in a glass wool plant between 1955 and 1977 followed to 1984 (97% traced; 155 deaths); comparison with age- and calendar-specific death rates of Ontario males	Deaths from NMRD fewer than expected; significant excess of lung cancer (O:E = 19:9.5) among "plant only" employees not related to duration of, or time since first, exposure	Data on smoking not available, but increase probably too large to be attributable solely to this cause; local mortality rates for lung cancer similar to those of Ontario and Canada; no historical exposure data	Shannon et al. (1984a,b, in press)
135 026 Swedish male construction workers examined by the Construction Industry Organization for Working Environment, Safety and Health between 1971 and 1974 and followed up to 1983 (99.9% traced; 7356 deaths); comparison of cancer mortality and incidence with age-specific national rates; smoking habits and place of residence taken into account in the analyses	Mortality from NMRD lower than expected; significant increase in cancer incidence for pleural mesothelioma (O:E = 23:10.8)	Possible selection bias ("survivor population"); exposure estimates based on job category and self-reported information only; construction workers frequently exposed to many dusts and difficulty in characterizing intermittent exposure	Engholm et al. (1984, in press)

Table 22 (contd).

Case-control study of 518 construction workers with respiratory cancer from above cohort compared with maximum of 5 controls each matched for age, time of first check-up, and survival; smoking habits and residence (urban/rural) taken into account in the analyses	Smoking and population adjusted relative risk for MMMF exposure (adjusted for asbestos exposure) - 1.21 and for asbestos exposure (adjusted for MMMF exposure) - 2.53	May be selection bias since possibly a "survivor population"; construction workers frequently exposed to many dusts and tremendous difficulty in characterizing exposure that was intermittent, varied in intensity, and based largely on worker recall	Engholm et al. (1984, in press)
1374 men with > 1 year employment in a glass wool factory between 1975 and 1984 followed up to 1984 (92.7% traced); comparison with age- and calendar-specific incidence for 3 regional French cancer directories (not including population in the region where the plant was located)	Significantly higher incidence of cancers of the larynx (SIR = 2.3), pharynx (SIR = 1.4), and buccal cavity (SIR = 3) in production but not in administration and maintenance workers; some relationship between SIRs and duration of exposure (not statistically significant)	Large loss to follow-up; those lost were considered to be alive; authors reported that mortality in region where plant located did not differ significantly from comparison regions, but did not indicate basis for the conclusion; it was also reported that tobacco smoking no more frequent in cohort than in general population (based on estimates in workers in 1983); mean time since first exposure, 17.6 years; number of expected lung cancers in workers with more than 20 years exposure small; study conducted due to observation by an industrial physician of an excess of cancers of the upper respiratory and alimentary tracts - essentially confirmation of a case report	Moulin et al. (1986)

[a] Excesses that were statistically signficant at $P < 0.05$ are described as "significant"; all other excesses were not statistically signficant.

NMRD = Non-malignant respiratory disease.
MRD = Malignant respiratory disease.
SMR = Standardized mortality ratio.
SIR = Standardized incidence ratio.

The most extensive historical prospective studies of disease incidence and mortality in populations occupationally exposed in the production of MMMF have been those conducted in the USA by Enterline et al. (in press) and in 7 European countries by Simonato et al. (1986a,b, in press). The results of studies on widely overlapping- or sub-cohorts of these 2 large investigations have also been reported (Enterline & Henderson, 1975; Bayliss et al., 1976; Robinson et al., 1982; Morgan et al., 1984; Andersen & Langmark, 1986; Bertazzi et al., 1986; Claude & Frentzel-Beyme, 1986; Gardner et al., 1986; Olsen et al., 1986; Teppo & Kojonen, 1986; Westerholm & Bolander, 1986). Discussion will be restricted mainly to results of: the complete cohorts in the 2 large investigations (Simonato et al., 1986a,b, in press; Enterline et al., in press), studies of two smaller, but distinct, cohorts of production workers (Moulin et al., 1986; Shannon et al., in press), and a cohort and case-control study of construction workers (Engholm et al., in press).

Standardized Mortality Ratios (SMRs) for non-malignant respiratory disease and their statistical significance at the conventional level of P = 0.05 for the main results in the European and US studies are presented in Table 23. In both studies, the authors preferred the use of local rates to national rates for the purposes of comparison and presented evidence of the reasons for this in fibrous glass workers (Gardner et al., 1986; Enterline et al., in press).

There has been little evidence of an excess of mortality from non-malignant respiratory disease (NMRD) in the cohort and case-control studies conducted to date. In the extensive European study of 21 967 production workers, there was no excess mortality from NMRD in the total cohort, in any production sector (i.e., continuous filament, glass wool, or rock wool/slag wool), or on analysis according to time from first exposure or duration of employment (Simonato et al., 1986a,b, in press). In the large cohort of 16 661 production workers in the USA, there was no excess mortality from NMRD in glass wool workers compared with local rates[a] (O = 144, SMR = 103), though there was a statistically significant excess compared with national rates (O = 158, SMR = 134). In mineral wool workers, there was a statistically nonsignificant excess of NMRD compared with both local (O = 31, SMR = 140) (Enterline et al., in press) and national rates (O = 31, SMR = 145) (Enterline & Marsh, in press). In continuous filament workers, there was no excess compared with either local (O = 35, SMR = 98) or national rates (O = 41, SMR = 90).

---

[a] In the US study, SMRs based on local rates are for the period 1960-82; SMRs based on national rates are for the period 1946-82.

Table 23. Non-malignant respiratory disease mortality - epidemiological studies of MMMF production workers[a]

Feature	Study	Fibre type		
		Glass filament	Glass wool[b,c]	Rock wool/ slag wool
Number of deaths from non-malignant respiratory disease	USA	35	144	31
	Europe	13	93	59
		Standardized mortality ratios compared with local rates[d,e]		
Non-malignant respiratory disease mortality	USA	98	102	140
	Europe (national)	96	105	94
Time since first exposure (< 10/ 10-19/20-29/ 30+ years)	USA	0/124/110/69	52/92/118/93	0/230/134/123
	Europe (national)	0/145/159/0	63/118/115/113	101/90/72/142
Duration of employment (< 20/20+ years) (> 20 years since first exposure)	USA	116/78		136/118
	Europe (national)	240/0	121/71	90/114
Estimated cumulative exposure (increasing intervals of fibre/cm^3· months)	USA	97/130/107/0	126/79/74/95	158/141/160/117
Small diameter fibres	USA			
- ever/never exposed		-	110/97	-
- by time since first exposure (ever exposed)		-	0/44/164/113	-

[a] IARC (1985) and Enterline & Marsh (in press).

[b] Data for "fibrous glass-both" and "fibrous glass-wool" plants in the USA study combined.

[c] In the only additional relevant study of a much smaller cohort of glass wool production workers, there was no excess of mortality from non-malignant respiratory disease (O = 4, SMR = 55) compared with provincial rates (Shannon et al., in press).

[d] Local rates, unless otherwise specified.

[e] No SMRs nor trends shown in this table are statistically significant at the conventional $P = 0.05$ level.

The results of analyses of NMRD mortality with time from first exposure, or duration of exposure, or cumulative fibre exposure did not showed any trends for either the glass wool or rock wool workers. Among glass wool workers "ever exposed" to small diameter fibres, there was no excess of NMRD mortality but a slight statistically insignificant increase with time since first exposure.

Shannon et al. (in press) studied 2557 glass wool workers with more than 90 days employment and found 4 deaths from NMRD, whereas 7.3 would have been expected (SMR = 55). Engholm et al. (in press), in their study of 135 026 Swedish construction workers potentially exposed to MMMF and asbestos, found 193 deaths from NMRD compared with 418 expected; this gave an unusually low SMR of 46.

8.1.3 Carcinogenicity

Standardized mortality ratios (SMRs) and their statistical significance at the conventional level of $P = 0.05$ for the main results in the European and US studies are presented in Table 24. In both studies, the authors preferred the use of local rates to national rates for the purposes of comparison and presented evidence of the reasons for this in fibrous glass workers (Gardner et al., 1986; Enterline et al., in press).

8.1.3.1 Glass wool

In the European study (Simonato et al., 1986a,b, in press), there was no significant excess of mortality from lung cancer (O = 93, SMR = 103) among glass wool production workers, when compared with local mortality rates. However, there was a statistically non-significant relationship between lung cancer mortality and time from first exposure. When compared with national mortality rates, there was a statistically significant excess of lung cancer that was related (O = 93, SMR = 127) (not statistically significantly) to time from first exposure. There was no relationship between lung cancer mortality and duration of employment. No excess was discernible among workers employed in the "early technological phase", whichever reference rate was used.

In the USA glass wool subcohort (Enterline et al., in press), the SMR for lung cancer was 109 (O = 267, statistically nonsignificant), based on local rates, and 116 (O = 267, statistically significant), based on national reference rates. After 20 years from the onset of exposure, the SMR was not statistically significant compared with local rates (111) but was statistically significant compared with US rates (124), based on 207 observed cases. There was a statistically nonsignificant increase with time since first exposure, but the

Table 24. Lung cancer mortality - epidemiological studies of MMMF production workers[a]

Feature	Study	Fibre type		
		Glass filament	Glass wool[b,c]	Rock/slag wool
Number of deaths from lung cancer	USA	64	267	60
	Europe	15	93	81
	Standardized mortality ratios compared with local rates			
Lung cancer mortality	USA	92	109	134 ($P < 0.05$)
	Europe	97	103	124
Time since first exposure (< 10/ 10-19/20-29/ 30+ years)	USA	104/53/119/80	92/108/108/114	90/157/127/135
	Europe	176/76/0/0	68/113/100/138	104/122/124/185
Duration of employment (< 20/20+ years) (> 20 years since first exposure)	USA	110/106[d]		145/111
	Europe	0/0	118/60	143/141
Estimated cumulative exposure (increasing intervals of fibre/cm^3· months)	USA	96/51/109/63	120/109/81/108	185/164/119/104
Technological phase	Europe			
- early/intermediate/late		-	92/111/77	257/141/111 ($P < 0.05$)[e]
- by time since first exposure (early phase)		-	108/70/80/121	0/0/317/295

trend was less evident when local rates were used. There was no relationship between respiratory cancer mortality and duration of exposure, or cumulative fibre exposure. There was also an excess (not statistically significant) of mortality due to lung cancer among 1015 workers "ever exposed" in the production of fibres of nominal diameter < 3 µm (O = 22, SMR = 124), where measured fibre levels were higher than in other production

Table 24 (contd).

Feature	Study	Fibre type		
		Glass filament	Glass wool[c,d]	Rock/slag wool
	Standardized mortality ratios compared with local rates			
Small diameter fibres	USA			
- ever/never exposed		-	124/105	-
- by time since first exposure (ever exposed)		-	61/128/105/198	-
Estimated concentrations of respirable fibres	USA	lower	intermediate (highest in small diameter fibre production facilities)	higher

[a] Modified from: IARC (in press).
[b] Data for "fibrous glass-both" and "fibrous glass-wool" plants in the USA study combined.
[c] In the only additional relevant study of a much smaller cohort of glass wool production workers, there was a statistically significant excess of lung cancer mortality compared with provincial rates (O = 19, SMR = 199), which was not related to time from first exposure (< 10 years, SMR = 241; 10+ years, SMR = 195) or duration of employment (< 5 years, SMR = 291; > 5 years, SMR = 174) (Shannon et al., in press).
[d] Data reported for "fibrous glass" (type unspecified) subcohort.
[e] Statistical test for linear trend.

sectors, compared with those "never exposed" in this sector. The excess showed an increasing but non-significant relationship with time from first exposure.

A case-control study nested in the US cohort was conducted by Enterline et al. (in press) with an initial number of 211 respiratory cancer cases among "fibrous glass" (type unspecified) workers and 374 controls. The logistic analysis performed indicated a statistically significant association between lung cancer and the smoking habits of workers but not between lung cancer and cumulative fibre dose.

In a glass wool plant in Ontario, there was a statistically significant (2-fold) excess of mortality from lung cancer, compared with provincial rates, among 2557 "plant only" employees with more than 90 days employment (O = 19, SMR = 199) (Shannon et al., in press). The lung cancer rates did not appear to be related to time from first exposure or duration of employment.

The rate of mesothelioma in glass wool production workers in the large US study was not excessive. Two cases were reported. No cases in glass wool workers were reported in the European study.

Moulin et al. (1986) reported a statistically significant higher incidence of cancers of the buccal cavity (O = 9, Standardized Incidence Ratio (SIR) = 301) in production workers with more than 1 year of employment in a glass wool factory in France; the excess was related to duration of exposure but not significantly so. This study was conducted following an observation by an industrial physician of an excess of cancers of the upper respiratory and alimentary tracts and must, therefore, be considered to be essentially a confirmation of a case report. On their own, therefore, these findings do not constitute convincing evidence of a real association. Moreover, the reported SIRs were based on comparison with rates of regions other than the one in which the plant was located.

A statistically nonsignificant excess of mortality due to laryngeal cancer was observed in a subcohort of the European study among 1098 Italian workers employed for at least 1 year in a plant manufacturing glass wool (Bertazzi et al., 1986). There was some relationship with time from first exposure, but the number of cases was small (O = 4, SMR = 190). These findings were limited to the Italian plant and were not consistent with either the results of the complete European or US studies. Moulin et al. (1986) also reported a statistically non-significant excess in the incidence of cancer of the larynx (O = 5, SIR = 230) in their cohort of 1374 glass wool production workers with more than 1 year of employment. However, this observation should be interpreted with caution, owing to the limitations of this study mentioned earlier.

8.1.3.2 Rock wool and slag wool

In the European subcohort of rock wool/slag wool production workers, there was a statistically nonsignificant excess of mortality from lung cancer, which was the same compared with either local or national mortality rates (O = 81, SMR = 124). There was a statistically nonsignificant relationship with time from first exposure (Simonato et al., 1986a,b, in press) but not with duration of employment. The excess was concentrated among workers employed in the "early technological phase" (with large SMRs of 257 and 214 for local and national comparisons, respectively, both statistically significant and based on 10 observed cases), in which airborne fibre levels were estimated to have been much higher. The use of copper slag containing arsenic was also reported in one factory during this period and ventilation was generally poor, possibly resulting in some exposure to polycyclic aromatic hydrocarbons (PAHs) from the furnaces. Other environmental contaminants that might have had an influence on the lung cancer mortality excess have also been analysed. A statistically significant increase in lung cancer mortality after 20 years since first exposure was associated with the use of slag (O = 23, SMR = 189). However, these findings are difficult to interpret due to the wide overlapping between the period of use of slag and the early technological phase. Neither the use of bitumen and pitch nor the presence of asbestos in some products explained the excess of lung cancer (Simonato et al., in press). The results of analysis of data on lung cancer incidence in the European cohort were similar to those reported for mortality.

Enterline et al. (in press) studied a cohort of 1846 white male workers from 6 US plants that produced slag wool or rock wool/slag wool. The SMRs for respiratory cancer were 134 (local rates) and 148 (national rates), both statistically significant, based on 60 observed cases. After 20 years since first exposure, the SMRs were 131 (local rates) and 146 (national rates), the latter being statistically significant, based on 45 observed cases. No clear trend with time since first exposure was found, and there was no relationship with duration of employment, or with estimated cumulative exposure in this subcohort of the US study. It was reported that one of the plants studied used a copper slag containing arsenic.

A case-control study nested in the US rock wool/slag wool production subcohort was also carried out using the same methodology as that for the glass wool production workers. The study comprised 45 respiratory cancer cases and 49 controls. Only the relationship between lung cancer and smoking was statistically significant ($P < 0.001$). In a further analysis, using 38 cases and 43 controls, there was a statistically significant relationship between lung cancer and estimated cumulative fibre dose ($P < 0.01$), after controlling for years of smoking, time since starting to smoke, and interaction terms between smoking and exposure. Reservations about the model used for this analysis complicate the interpretation of these results.

Only one case of pleural mesothelioma was reported in the rock wool/slag wool production workers in the US cohort. Similarly, in the large European cohort (Simonato et al., 1986a,b, in press), only one pleural mesothelioma occurred in a rock wool/slag wool production plant with a brief latency period following a relatively short exposure time. This single case does not represent an excess.

In the European study, there was a statistically significant excess in the incidence of cancer of the buccal cavity and pharynx (O = 22, SIR = 180), which showed a statistically nonsignificant relationship with time from first exposure. Also observed in the European study was a statistically non-significant excess in bladder cancer mortality (O = 13, SMR = 137) in the rock wool/slag wool production sectors, which was related at a statistically significant level to time from first exposure (IARC, 1985). In neither of these cases was there an association with the early technological phase when fibre levels were estimated to be higher.

8.1.3.3 Glass filament

In the glass filament subcohort of the European study, there was no excess of mortality from lung cancer compared with local mortality rates (O = 15, SMR = 97), but a small excess (not statistically significant) compared with national rates (O = 15, SMR = 120), which was not related to time from first exposure (Simonato et al., 1986a,b, in press). However, it should be noted that the length of follow-up of the continuous filament workers in this study was short.

No excess for lung cancer was reported by Enterline et al. (in press) from the follow-up of 3435 white male workers employed in glass filament production in the large US study. The SMRs for respiratory cancer were 92 (local rates) and 95 (national rates), based on 64 observed cases. There was no relationship with time since first exposure, or with estimated cumulative fibre dose.

8.1.3.4 Mixed exposures

Engholm et al. (in press) reported the results of the follow-up of a large cohort of construction workers in Sweden. The SIR for lung cancer was 91 based on 440 observed cases. The authors also investigated the possible influence of exposure to asbestos and to MMMF using a nested case-referent approach. After adjusting for asbestos exposure, the relative risk for exposure to MMMF was 1.21 (95% CI: 0.60 - 2.47), while it was 2.53 (95% CI: 0.77 - 8.32) for asbestos after controlling for potential MMMF exposure. A wide overlapping of asbestos and MMMF exposures makes the analysis and interpretation of the results of this study difficult.

8.1.3.5 Refractory fibres

No published epidemiological studies are available.

8.2 General Population

With the exclusion of isolated case reports of respiratory symptoms and dermatitis associated with exposure to MMMF in the home and office environments, and 2 limited cross-sectional studies of ocular and respiratory effects in offices and schools, adverse effects on the general population have not been reported. For example, Newball & Brahim (1976) attributed respiratory symptoms in members of a family to fibrous glass exposure from a residential air conditioning system. Dermatitis has been observed in individuals exposed to disturbed MMMF insulation in office buildings (Verbeck et al., 1981; Farkas, 1983) and to clothing contaminated during laundering with MMMF-containing materials (Lucas, 1976).

A significant increase in the incidence of "smarting" and watering eyes, swollen eyelids, disturbance of sight, conjunctival hyperaemia, and "smarting" of the nose was observed in 39 persons working in buildings with ceilings made of compressed mineral wool compared with a control group of 23 persons, matched for age, sex, and smoking habits, who were working in buildings with plaster ceilings. In 5 of the exposed individuals, mineral fibres were present in the conjunctival mucous threads compared with 0 in the control group. It was also reported that the occurrence of symptoms and conjunctival hyperaemia were significantly reduced by surface treatment of the mineral wool ceiling (Alsbirk et al., 1983).

The association between various signs and symptoms of disease and exposure to MMMF in adults and children in 24 kindergartens was investigated in one Danish county in Autumn 1984 (Rindel et al., 1987). In 10 kindergartens, the ceilings were covered with MMMF products with water-soluble binder (Group A), 6 had resin-bound ceiling covering material (Group B), and 8 had ceilings on which MMMF was not apparent (Group C). The investigation included questionnaires for the adults (n = 200, 92% response) concerning health and socio-economic status, smoking habits, and contact lens use (mode of administration not stated); questionnaires for children (n = 900, 90% response) were completed by their parents. In addition, for 3 months, daily records were kept by the adults of their own signs and symptoms of disease and those of the children. All subjects were also clinically examined by a doctor. Concentrations of fibres and dust, carbon dioxide levels, temperature, humidity, and air speed were also determined. Mean total airborne MMMF levels were similar in Groups A and B (23 fibres/m^3 and 40 fibres/m^3, respectively), but ranged from 0 to 77 fibres/m^3 in Group C. Among children, there was no correlation between disease symptoms and MMMF concentrations. For adults, eye irritation was significantly related to respirable ($P = 0.03$) and non-respirable ($P = 0.004$) MMMF. The presence of settled MMMF on surfaces was correlated with adult skin irritation ($P = 0.005$). However, the authors concluded that the total MMMF in air derived from ceiling coverings did not explain the reported symptoms and diseases. Unless the adults were not informed of the hypothesis to be tested, and the physician involved "blinded", it is difficult to see how bias could have been avoided.

There have not been any mortality or cancer morbidity studies concerning exposure to MMMF in the general population.

9. EVALUATION OF HUMAN HEALTH RISKS

Whenever possible, emphasis is placed on data obtained in epidemiological studies of human populations exposed to MMMF, though few data on quantitative exposure-response relationships are available. Information from animal and *in vitro* studies is used mainly in the comparative assessment of the potency of various fibre types and sizes, particularly when human epidemiological studies are lacking or the results are not clear cut, as is the case for some MMMF.

9.1 Occupationally Exposed Populations

There have been isolated reports of dermatitis and eye irritation in workers exposed to MMMF (section 8.1.1). However, no human data are available concerning the exposure-response relationship for these effects, and animal studies have not been conducted to evaluate either of them specifically.

In addition, there has been some suggestion of non-malignant respiratory effects (e.g., decrements in lung function) in MMMF-exposed workers (section 8.1.2), but some of these studies have been methodologically weak. In the study regarded as the best to date, a slight excess of small radiological opacities was observed, but this was not accompanied by ventilatory decrement or increase in respiratory symptoms. The results of experimental animal studies have indicated that MMMF might be potentially fibrogenic, especially when introduced into body cavities (pleural and peritoneal) (section 7.1.1.1). However, a significant fibrogenic response has not been seen in inhalation studies, to date, though it should be noted that the information in this regard is more complete for glass fibres than for other MMMF. Thus, it is not possible, on the basis of available data from human epidemiological and experimental animal studies, to draw definite conclusions concerning the nature and extent of the possible non-malignant hazards for the respiratory system resulting from exposure to MMMF.

An important concern is the potential risk of cancer in workers exposed to MMMF. Although there is no evidence that pleural or peritoneal mesotheliomas have been associated with occupational exposure in the production of various MMMF, there have been indications of increases in lung cancer mortality from the main epidemiological studies of workers in some sectors of MMMF production, including the two most extensive investigations of workers in Europe and the USA (section 8.1.3).

In the rock wool/slag wool production industry, the standardized mortality ratios for lung cancer in the large European and US cohorts were 124 and 134, respectively (Simonato et al., 1986a,b, in press; Enterline et al., in press). In the glass wool production industry, the corresponding SMRs were 103 and 109, respectively. There has been no increase in lung cancer mortality in continuous filament production workers; SMRs in the European and US studies were 97 and 92, respectively. However, available data are too few to establish a quantitative exposure-response relationship for lung cancer mortality in relation to airborne fibre concentrations.

There has been some suggestion in these cohort studies of increases in cancer at sites other than the lung, but, it is not possible, on the basis of the available data, to draw any conclusions concerning the possible role of occupational exposure during MMMF production in the etiology of these other malignant diseases.

It has been suggested that other factors present in the work-place may have contributed to the increased lung cancer mortality observed, such as the presence of contaminants in the slag used, particularly in the early phase of mineral wool production in the European study. However, where it has been possible to study such contaminants, including asbestos and arsenic in copper slag, the lung cancer excess was not explained by their presence. Furthermore, it is not likely that any known potential confounding factors for lung cancer mortality, including cigarette smoking, could explain the high rates observed in the early technological phase of the rock wool/slag wool industry in Europe. The hypothesis that the airborne fibre concentrations are the most important determinants of lung cancer risk is supported by the observation of a qualitative relationship between the standardized mortality ratios and past estimated airborne fibre levels in the various sectors (rock wool/slag wool, glass wool, and continuous filament) of the production industry. Moreover, while not statistically significant, there was a rise in lung cancer risk that increased with length of time since first exposure in workers involved in the production of smaller diameter (< 3 μm) glass wool fibres in the USA. Airborne fibre levels measured in this sector are higher than those in the rest of the glass wool production sector. However, it should be emphasized that, at the lower fibre concentrations associated with the improved production conditions of the late technological phase, no excess of lung cancer mortality has been observed in the European rock wool/slag wool workers.

Airborne MMMF concentrations present in work-places with good work practices are generally of the order of, or less than, 0.1 fibres/cm^3 (section 5). However, data reviewed in section 5 indicate that mean airborne fibre levels for some workers in the ceramic fibre and small diameter (< 1 μm) glass wool fibre manufacturing sectors may be similar to those to which workers were exposed in the early production phase. Therefore, although only a small proportion of workers are employed in these segments of the industry, their lung cancer risk could potentially be elevated. However, epidemiological data are not yet available on workers in the ceramic fibre industry. Elevated average concentrations of fibres during the blowing or spraying of insulation wool in confined spaces have also been recorded and, though the time-weighted average exposure for individual workers may not be as high (section 5), their lung cancer risk could similarly be increased, if protective equipment is not used.

The results of animal studies, *in toto,* indicate that the excess of lung cancer observed in the epidemiological studies in some sectors of the MMMF production industry is biologically plausible (section 7.1). Although in inhalation studies (probably the most relevant studies for risk assessment in man) conducted to date, MMMF have not induced a significant carcinogenic response, the carcinogenic potential of many MMMF has been demonstrated when they have been introduced directly into the pleural and peritoneal cavities. The results of these studies clearly show that the carcinogenic potential in animals is directly related to fibre size and durability. In addition, glass fibres have been shown to cause chromosomal alterations and cell transformation *in vitro,* but little information in this regard is available for the other MMMF (section 7.2).

9.2 General Population

There have been isolated case reports of respiratory symptoms and dermatitis associated with exposure to MMMF in the home and office environments (section 8.2). However, available epidemiological data are not sufficient to draw any conclusions in this respect. As with occupationally-exposed populations, it is the potential risk of lung cancer at low levels of exposure that is of most concern, but no direct evidence is available from which to draw conclusions.

Available data on occupationally-exposed populations are not yet sufficient to estimate quantitatively, by extrapolation to low levels, the risk of lung cancer in the general population associated with exposure to MMMF in the environment. Moreover, the available information is inadequate to characterize exposure to MMMF in the general environment. However, levels of MMMF in the typical indoor and general environments, measured to date, are very low compared with present levels in most sectors of the production and user industry and certainly much lower (by several orders of magnitude) (sections 5.1.1.2 and 5.1.1.3) than some past occupational exposure levels associated with raised lung cancer risks. It should also be noted that such increases in lung cancer risk have not been observed among workers employed under the improved conditions of the late technological phase and followed up for a sufficient length of time.

The overall picture indicates that the possible risk of lung cancer among the general public is very low, if there is any at all, and should not be a cause for concern if the current low levels of exposure continue.

10. RECOMMENDATIONS

10.1 Further Research Needs

10.1.1 Analytical methods

The current reference scheme for the measurement of MMMF concentrations in the occupational environment by the membrane filter optical and scanning electron microscopic methods should continue.

There is also a need to standardize methods of measurement of MMMF in ambient and indoor air; ideally, such methods should determine both the total mass of dust and the number of fibres likely to be deposited in the respiratory system. Where scanning electron or transmission electron microscopy are used, fibre size distribution should be fully characterized. Greater standardization of sample preparation and measurement methods for the determination of MMMF in other media, including biological tissues, is also desirable.

Further research on the development of automated fibre counting methods, which may improve the consistency of results obtained in different laboratories, should be conducted. The development of cheaper and more practical methods of determining fibrous particulates is also desirable.

For fibrous dusts with the potential for causing non-malignant and malignant diseases of the airways and lung parenchyma, attention should, in future, be paid to the measurement and study of the effects of the fraction of fibres (inhalable) that is mainly deposited on the surface of the airways as well as the fraction ("respirable") with a high efficiency for airway penetration and deposition in the alveoli.

10.1.2 Environmental exposure levels

There is a need to determine levels of MMMF in the environment that are representative of general population exposure, and it is recommended that respirable levels of MMMF in ambient and indoor air should be measured by appropriate and preferably standardized methods of sampling and analysis. Analysis of the MMMF contents of tissues in both occupationally exposed groups and the general population is also recommended.

10.1.3 Studies on animals

Standardized reference samples of MMMF products as well as size-selected fibre samples should be developed for use in experimental animal studies. Complete characterization and reporting of the fibre size distributions of MMMF used in such investigations is also essential.

Research to investigate the relative potencies of the various MMMF types (preferably by routes of exposure that simulate those of man, e.g., inhalation) should continue. The effects of fibre coating on the potential of MMMF to cause disease should be further examined and investigations of the effects of concomitant exposure to MMMF and other airborne pollutants (such as tobacco smoke) should be conducted.

Studies should be undertaken to examine further the mechanisms by which fibrous materials cause disease. Particularly relevant for the assessment of risks associated with exposure to the MMMF is the examination of the biological significance of the durability of fibres in tissue. Development of short-term models of durability are needed, as well as a comparison of durability in different locations in the body.

Studies on the relationship of fibrogenicity and carcinogenicity are of special importance as is a better understanding of the mechanisms of fibre toxicity, especially at the cellular level (interaction of macrophages, etc.). Cytotoxicity and genotoxicity studies should be extended to all MMMF (currently available only on glass fibres).

As new fibres are developed, they should be subjected to systematic investigation for the assessment of health hazard. It would be desirable to develop a standard set of study protocols for use in this regard.

10.1.4 Studies on man

Further study of the prevalence of dermatitis and eye irritation in populations exposed occupationally to MMMF is warranted. The respiratory effects observed in cross-sectional studies of MMMF workers should be further investigated using standardized methods to measure effects, including the International Classification of Radiographs of Pneumoconiosis, in appropriately designed epidemiological studies (WHO, 1983b).

Follow-up of MMMF-exposed workers examined in the historical prospective studies reported to date should continue. The possibilities of extending these investigations to meaningful studies (in terms, for example, of size and time since first exposure) on workers who have entered the glass wool and rock wool industries more recently, those working with refractory (including ceramic) fibres, special purpose fibres, or other new fibres, and workers in the user industries should be explored. These studies should be accompanied by the collection of agreed comprehensive industrial hygiene data in advance of the epidemiological analyses. Where possible, continued attempts should be made to control for confounding factors, such as cigarette smoking (perhaps by nested case-referent studies) and past exposure to asbestos (perhaps by lung fibre burden studies).

10.2 Other Recommendations

10.2.1 Classification of MMMF products

There is a need for setting up a systematic scheme for the classification of the various MMMF products manufactured.

There is also a need to give guidance concerning the potential for fibre release from MMMF products. As a first step, information on fibre diameters in the bulk material is essential.

REFERENCES

AALTO, M. & HEPPLESTON, A.G. (1984) Fibrogenesis by mineral fibres: an *in vitro* study of the roles of the macrophage and fibre length. *Br. J. exp. Pathol.*, **65**: 91-99.

ALSBIRK, K.E., JOHANSSON, M., & PETERSEN, R. (1983) [Ocular symptoms and exposure to mineral fibres in boards for sound insulation of ceilings.] *Ugeskr. Laeger*, **145**: 43-47 (in Danish with English summary).

ANDERSEN, A. & LANGMARK, F. (1986) Incidence of cancer in the mineral-wool producing industry in Norway. *Scand. J. Work environ. Health*, **12**(Suppl. 1): 72-77.

ARBOSTI, G., LO MARTIRE, N., & BONARI, R. (1980) [Dermatological and allergic pathology in workers of a glass fibre factory.] *Med. Lav.*, **1**: 99-105 (in Italian with English abstract).

BALZER, J.L. (1976) Environmental data: airborne concentrations found in various operations. In: *Occupational Exposure to Fibrous Glass. Proceedings of a Symposium, College Park, Maryland, 26-27 June 1974*, Washington DC, US Department of Health, Education and Welfare, pp. 83-89.

BALZER, J.L., COOPER, W.C., & FOWLER, D.P. (1971) Fibrous glass-lined air transmission systems: an assessment of their environmental effects. *Am. Ind. Hyg. Assoc. J.*, **32**: 512-518.

BAYLISS, D.L., DEMENT, J.M., WAGONER, J.K., & BLEJER, H.P. (1976) Mortality patterns among fibrous glass production workers. *Ann. NY Acad. Sci.*, **271**: 324-335.

BECK, E.G. (1976a) [The interaction between cells and fibrous dusts.] *Zbl. Bakteriol. Hyg. I. Abt. Orig. B*, **162**: 85-92 (in German with English abstract).

BECK, E.G. (1976b) Interaction between fibrous dust and cells *in vitro*. *Ann. Anat. Pathol.*, **12**(2):227-236

BECK, E.G. & BRUCH, J. (1974) [Effet des poussières fibreuses sur les macrophages alvéolaires et sur d'autres cellules cultivées *in vitro*. Etude biochimique et morphologique.] *Rev. fr. Mal. respir.*, **2**(Suppl. 1): 72-76 (with English summary).

BECK, E.G., HOLT, P.F., & MANOJLOVIC, N. (1972) Comparison of effects on macrophage cultures of glass fibre, glass powder, and chrysotile asbestos. *Br. J. ind. Med.*, **29**: 280-286.

BELLMANN, B., MUHLE, H., POTT, F., KONIG, H., KLOPPEL, H., & SPURNY, K. (in press) Persistence of man-made mineral fibres and asbestos in rat lungs. *Ann. occup. Hyg.* **31**(4B).

BERNSTEIN, D.M., DREW, R.T., & KUSCHNER, M. (1980) Experimental approaches for exposure to sized glass fibers. *Environ. Health Perspect.*, **34**: 47-57.

BERNSTEIN, D.M., DREW, R.T., SCHIDLOVSKY, G., & KUSCHNER, M. (1984) Pathogenicity of MMMF and the contrasts with natural fibres. In: *Biological Effects of Man-Made Mineral Fibres. Proceedings of a WHO/IARC Conference, Copenhagen, Denmark, 20-22 April 1982*, Copenhagen, World Health Organization, Regional Office for Europe, **Vol. 2**, pp. 169-195.

BERTAZZI, P.A., ZOCCHETTI, C., RIBOLDI, L., PESATORI, A., RADICE, L., & LATOCCA, R. (1986) Cancer mortality of an Italian cohort of workers in man-made glass-fibre production. *Scand. J. Work environ. Health*, **12**(Suppl. 1): 65-71.

BIGNON, J., MONCHAUX, G., CHAMEAUD, J., JAURAND, M.C., LAFUMA, J., & MASSE, R. (1983) Incidence of various types of thoracic malignancy induced in rats by intrapleural injection of 2 mg of various mineral dusts after inhalation of ^{222}Ra. *Carcinogenesis*, **4**: 621-628.

BISHOP, K., RING, S.J., ZOLTAI, T., MANOS, C.G., AHRENS, V.D., & LISK, D.J. (1985) Identification of asbestos and glass fibres in municipal sewage sludges. *Bull. environ. Contam. Toxicol.*, **34**: 301-308.

BJORNBERG, A. (1985) Glass fiber dermatitis. *Am. J. ind. Med.*, **8**: 395-400.

BJORNBERG, A. & LOWHAGEN, G. (1977) Patch testing with mineral wool (rockwool). *Acta dermatovenerol. (Stockholm)*, **57**: 257-260.

BJORNBERG, A., LOWHAGEN, G., & TENGBERG, J-E. (1979) Relationship between intensities of skin test reactions to glass-fibres and chemical irritants. *Contact Dermatit.*, **5**: 171-174.

BROWN, R.C., CHAMBERLAIN, M., & SKIDMORE, J.W. (1979a) *In vitro* effects of man-made mineral fibres. *Ann. occup. Hyg.*, **22**: 175-179.

BROWN, R.C., CHAMBERLAIN, M., DAVIES, R., GAFFEN, J., & SKIDMORE, J.W. (1979b) *In vitro* biological effects of glass fibers. *J. environ. Pathol. Toxicol.*, **2**: 1369-1383.

BROWN, G.M., COWIE, H., DAVIS, J.M.G., & DONALDSON, K. (1986) *In vitro* assays for detecting carcinogenic mineral fibres: a comparison of two assays and the role of fibre size. *Carcinogenesis*, **7**(12): 1971-1974.

BRUCH, J. (1974) Response of cell cultures to asbestos fibers. *Environ. Health Perspect.*, **9**: 253-254.

BURDETT, G.J. & ROOD, A.P. (1983) Membrane-filter, direct-transfer technique for the analysis of asbestos fibers or other inorganic particles by transmission electron microscopy. *Environ. Sci. Technol.*, **17**(11): 643-648.

BURDETT, G.J., KENNY, L.C., OGDEN, T.L., ROOD, A.P., SHENTON-TAYLOR, T., TARRY, R., & VAUGHAN, N.P. (1984) Problems of fibre counting and its automation. In: *Biological Effects of Man-Made Mineral Fibres. Proceedings of a WHO/IARC Conference, Copenhagen, Denmark, 20-22 April 1982*, Copenhagen, World Health Organization, Regional Office for Europe, **Vol. 1**, pp. 201-216.

CARPENTER, J.L. & SPOLYAR, L.W. (1945) Negative chest findings in a mineral wool industry. *J. Indiana State Med. Assoc.*, **38**: 389-390.

CASEY, G. (1983) Sister-chromatid exchange and cell kinetics in CHO-K1 cells, human fibroblasts and lymphoblastoid cells exposed *in vitro* to asbestos and glass fibre. *Mutat. Res.*, **116**: 369-377.

CHAMBERLAIN, M. & TARMY, E.M. (1977) Asbestos and glass fibres in bacterial mutation tests. *Mutat. Res.*, **43**: 159-164.

CHATFIELD, E.R. (1983) *Measurement of asbestos fibre concentrations in ambient atmospheres*, Toronto, Royal Commission on Matters of Health and Safety Arising from the Use of Asbestos in Ontario, 117 pp (Study No. 10).

CHERRIE, J. & DODGSON, J. (1986) Past exposures to airborne fibers and other potential risk factors in the European man-made mineral fibres production industry. *Scand. J. Work environ. Health*, **12**(Suppl. 1): 26-33.

CHERRIE, J., DODGSON, J., GROAT, S., & MACLAREN, W. (1986) Environmental surveys in the European man-made mineral fibre production industry. *Scand. J. Work environ. Health*, **12**: 18-25.

CHERRIE, J., KRANTZ, S., SCHNEIDER, T., OHBERG, I., KAMSTRUP, O., & LINANDER, W. (in press) An experimental simulation of an early rockwool/slagwool production process. *Ann. occup. Hyg.*

CHOLAK, J. & SCHAFER, L.J. (1971) Erosion of fibers from installed fibrous-glass ducts. *Arch. environ. Health*, **22**: 220-229.

CLAUDE, J. & FRENTZEL-BEYME, R.R. (1986) Mortality of workers in a German rock-wool factory - a second look with extended follow-up. *Scand. J. Work environ. Health*, **12**(Suppl. 1): 53-60.

CONDE-SALAZAR, L., GUIMARAENS, D., ROMERO, L.V., HARTO, A., & GONZALEZ, M. (1985) Occupational dermatitis from glass fiber. *Contact Dermatit.*, **13**: 195-196.

CORN, M. (1979) An overview of inorganic man-made fibers in man's environment. In: Lemen, R. & Dement, J.M., ed. *Dusts and disease*, Park Forest South, Illinois, Pathotox Publishers, pp. 23-36.

CORN, M. & SANSONE, E.B. (1974) Determination of total suspended particulate matter and airborne fiber concentrations at three fibrous glass manufacturing facilities. *Environ. Res.*, **8**: 37-52.

CORN, M., HAMMAD, Y., WHITTIER, D., & KOTSKO, N. (1976) Employee exposure to airborne fiber and total particulate matter in two mineral wool facilities. *Environ. Res.*, **12**: 59-74.

CRAWFORD, N.P., KELLO, D., & JARVISALO, J.O. (in press) Monitoring and evaluating man-made mineral fibres: work of a WHO/EURO reference scheme. *Ann. occup. Hyg.* **31**(4B).

CUYPERS, J.M.C., BLEUMICK, E., & NATER, J.P. (1975) [Dermatological aspects of glass fibre manufacturing,] **23**: 143-154 (in German).

DAVIES, R. (1980) The effect of mineral fibres on macrophages. In: Wagner, J.C., ed. *Biological effects of mineral fibres*, Lyons, International Agency for Research on Cancer, **Vol. 1**, pp. 419-425 (IARC Scientific Publication 30).

DAVIS, J.M.G. (1972) The fibrogenic effects of mineral dusts injected into the pleural cavity of mice. *Br. J. exp. Pathol.*, **53**: 190-201.

DAVIS, J.M.G. (1976) Pathological aspects of the injection of glass fiber into the pleural and peritoneal cavities of rats and mice. In: *Occupational Exposure to Fibrous Glass. Proceedings of a Symposium, College Park, Maryland, 26-27 June 1974*, Washington DC, US Department of Health, Education and Welfare, pp. 141-150.

DAVIS, J.M.G. (1981) The biological effects of mineral fibres. *Ann. occup. Hyg.*, **24**: 227-230.

DAVIS, J.M.G., GROSS, P., & DE TREVILLE, R.T.P. (1970) "Ferruginous bodies" in guinea-pigs. Fine structure produced experimentally from minerals other than asbestos. *Arch. Pathol.*, **89**: 364-373.

DAVIS, J.M.G., ADDISON, J., BOLTON, R.E., DONALDSON, K., JONES, A.D., & WRIGHT, A. (1984) The pathogenic effects of fibrous ceramic aluminum silicate glass administered to rats by inhalation or peritoneal injection. In: *Biological Effects of Man-Made Mineral Fibres. Proceedings of a WHO/IARC Conference, Copenhagen, Denmark, 20-22 April 1982*, Copenhagen, World Health Organization, Regional Office for Europe, **Vol. 2**, pp. 303-322.

DAVIS, J.M.G., GLYSETH, B., & MORGAN, A. (1986) Assessment of mineral fibres from human lung tissue. *Thorax*, **41**: 167-175.

DEMENT, J.M. (1975) Environmental aspects of fibrous glass production and utilization. *Environ. Res.*, **9**: 295-312.

DENIZEAU, F., MARION, M., CHEVALIER, G., & COTE, M.G. (1985) Ultrastructural study of mineral fiber uptake by hepatocytes (attapulgite, xonotlite, sepiolite, isolated liver cells, phagocytosis) *in vitro*. *Toxicol. Lett.*, **26**: 119-126.

DODGSON, J., CHERRIE, J.W., & GROAT, S. (in press) Estimates of past exposure to respirable fibres in the European man/made mineral fibre industry. *Ann. occup. Hyg.*

DREW, R.T., KUSCHNER, M., & BERNSTEIN, D.M. (in press) The chronic effects of exposure of rats to sized glass fibres. *Ann. occup. Hyg.* **31**(4B).

DUMAS, L. & PAGE, M. (1986) Growth changes of 3T3 cells in the presence of mineral fibers. *Environ. Res.*, **39**: 199-204.

ENGELBRECHT, F.M. & BURGER, B.F. (1975) Mesothelial reation to asbestos and other irritants after intraperitoneal injection. *S. Afr. med. J.*, **49**: 87-90.

ENGHOLM, G. & VON SCHMALENSEE, G. (1982) Bronchitis and exposure to man-made mineral fibres in non-smoking construction workers. *Eur. J. respir. Dis.*, **63**: 73-78.

ENGHOLM, G., ENGLUND, A., HALLIN, N., & SCHMALENSEE, G.V. (1984) Incidence of respiratory cancer in Swedish construction workers. In: *Biological Effects of Man-Made Mineral Fibres. Proceedings of a WHO/IARC Conference, Copenhagen, Denmark, 20-22*

April 1982, Copenhagen, World Health Organization, Regional Office for Europe, **Vol. 1**, pp. 350-366.

ENGHOLM, G., ENGLUND, A., FLETCHER, T., & HALLIN, N. (in press) Respiratory cancer incidence in Swedish construction workers exposed to man-made mineral fibres. *Ann. occup. Hyg.* **31**(4B).

ENTERLINE, P.E. & HENDERSON, V. (1975) The health of retired fibrous glass workers. *Arch. environ. Health,* **30**: 113-116.

ENTERLINE, P.E. & MARSH, G.M. (1984) The health of workers in the MMMF industry. In: *Biological Effects of Man-Made Mineral Fibres. Proceedings of a WHO/IARC Conference, Copenhagen, Denmark, 20-22 April 1982,* Copenhagen, World Health Organization, Regional Office for Europe, **Vol. 1**, pp. 311-339.

ENTERLINE, P.E. & MARSH, G.M. (in press) *A report to TIMA. Mortality among MMMF workers in the US: mortality update 1978-82,* Pittsburgh, Pennsylvania, Graduate School of Public Health, University of Pittsburgh.

ENTERLINE, P.E., MARSH, G.M., & ESMEN, N.A. (1983) Respiratory disease among workers exposed to man-made mineral fibers. *Am. Rev. respir. Dis.,* **128**: 1-7.

ENTERLINE, P.E., MARSH, G.M., HENDERSON, V., & CALLAHAN, C. (in press) Mortality update of a cohort of US man-made mineral fibre workers. *Ann. occup. Hyg.* **31**(4B).

ERNST, P., SHAPIRO, S., DALES, R.E., & BECKLAKE, M.R. (1987) Determinants of respiratory symptoms in insulation workers exposed to asbestos and synthetic mineral fibres. *Br. J. ind. Med.,* **44**: 90-95.

ESMEN, N.A. (1984) Short-term survey of airborne fibres in US manufacturing plants. In: *Biological Effects of Man-Made Mineral Fibres. Proceedings of a WHO/IARC Conference, Copenhagen, Denmark, 20-22 April 1982,* Copenhagen, World Health Organization, Regional Office for Europe, **Vol. 1**, pp. 65-82.

ESMEN, N.A., HAMMAD, Y.Y., CORN, M., WHITTIER, D., KOTSKO, N., HALLER, M., & KAHN, A. (1978) Exposure of employees to man-made mineral fibres: mineral wool production. *Environ. Res.,* **15**: 262-277.

ESMEN, N., CORN, M., HAMMAD, Y., WHITTIER, D., & KOTSKO, N. (1979a) Summary of measurements of employee exposure to airborne dust and fiber in sixteen facilities producing man-made mineral fibers. *Am. Ind. Hyg. Assoc. J.,* **40**: 108-117.

ESMEN, N.A., CORN, M., HAMMAD, Y.Y., WHITTIER, D., KOTSKO, N., HALLER, M., & KAHN, R.A. (1979b) Exposure of employees to man-made mineral fibers: ceramic fiber production. *Environ. Res.*, **19**: 265-278.

ESMEN, N.A., WHITTIER, D., KAHN, R.A., LEE, T.C., SHEEHAN, M., & KOTSKO, N. (1980) Entrainment of fibers from air filters. *Environ. Res.*, **22**: 450-465.

ESMEN, N.A., SHEEHAN, M.J., CORN, M., ENGEL, M., & KOTSKO, N. (1982) Exposure of employees to man-made vitreous fibers: installation of insulation materials. *Environ. Res.*, **28**: 386-398.

FARKAS, J. (1983) Fibreglass dermatitis in employees of a project-office in a new building. *Contact Dermatit.*, **9**: 79.

FERON, V.J., SCHERRENBERG, P.M., IMMEL, H.R., & SPIT, B.J. (1985) Pulmonary response of hamsters to fibrous glass: chronic effects of repeated intratracheal instillation with or without benzo(a)pyrene. *Carcinogenesis*, **6**: 1495-1499.

FINNEGAN, M.J., PICKERING, C.A.C., BURGE, P.S., GOFFE, T.R.P., AUSTWICK, P.K.C., & DAVIES, P. S. (1985) Occupational asthma in a fibre glass works. *J. Soc. Occup. Med.*, **35**: 121-127.

FISHER, A.A. (1982) Fiberglass vs mineral wool (rockwool) dermatitis. *Cutis*, **29**: 412-427.

FORGET, G., LACROIX, M.J., BROWN, R.C., EVANS, P.H., & SIROIS, P. (1986) Response of perifused alveolar macrophages to glass fibers: effect of exposure duration and fiber length. *Environ. Res.*, **39**: 124-135.

FORSTER, H. (1984) The behaviour of mineral fibres in physiological solutions. In: *Biological Effects of Man-Made Mineral Fibres. Proceedings of a WHO/IARC Conference, Copenhagen, Denmark, 20-22 April 1982*, Copenhagen, World Health Organization, Regional Office for Europe, **Vol. 2**, pp. 27-59.

FOWLER, D.P., BALZER, J.L., & COOPER, W.C. (1971) Exposure of insulation workers to airborne fibrous glass. *Am. Ind. Hyg. Assoc. J.*, **32**: 86-91.

FRIEDBERG, K.D. & ULLMER, S. (1984) Studies on the elimination of dust of MMMF from the rat lung. In: *Biological Effects of Man-Made Mineral Fibres. Proceedings of a WHO/IARC Conference, Copenhagen, Denmark, 20-22 April 1982*, Copenhagen, World Health Organization, Regional Office for Europe, **Vol. 2**, pp. 18-26.

GANTNER, B.A. (1986) Respiratory hazard from removal of ceramic fiber insulation from high temperature industrial furnaces. *Am. Ind. Hyg. Assoc. J.*, **47**: 530-534.

GARDNER, M.J., WINTER, P.D., PANNETT, B., SIMPSON, M.J.C., HAMILTON, C., & ACHESON, E.D. (1986) Mortality study of workers in the man-made mineral fiber production industry in the United Kingdom. *Scand. J. Work environ. Health*, **12**(Suppl. 1): 85-93.

GOLDSTEIN, B., RENDALL, R.E.G., & WEBSTER, I. (1983) A comparison of the effects of exposure of baboons to crocidolite and fibrous-glass dusts. *Environ. Res.*, **32**: 344-359.

GOLDSTEIN, B., WEBSTER, I., & RENDALL, R.E.G. (1984) Changes produced by the inhalation of glass fibre in non-human primates. In: *Biological Effects of Man-Made Mineral Fibres. Proceedings of a WHO/IARC Conference, Copenhagen, Denmark, 20-22 April 1982*, Copenhagen, World Health Organization, Regional Office for Europe, **Vol. 2**, pp. 273-285.

GOODGLICK, L.A. & KANE, A.B. (1986) Role of reactive oxygen metabolites in crocidolite asbestos toxicity to mouse macrophages. *Cancer Res.*, **46**: 5558-5566.

GRIFFIS, L.C., HENDERSON, T.R., & PICKRELL, J.A. (1981) A method for determining glass in rat lung after exposure to a glass fiber aerosol. *Am. Ind. Hyg. Assoc. J.*, **42**: 566-569.

GRIFFIS, L.C., PICKRELL, J.A., CARPENTER, R.L., WOLFF, R.K., MCALLEN, S.J., & YERKES, K.L. (1983) Deposition of crocidolite asbestos and glass microfibers inhaled by the beagle dog. *Am. Ind. Hyg. Assoc. J.*, **44**: 216-222.

GROSS, P. & BRAUN, D.C. (1984) *Toxic and biomedical effects of fibers*, Park Ridge, New Jersey, Noyes Publications, 257 pp.

GROSS, P., KASCHAK, M., TOLKER, E.B., BABYAK, M., & DE TREVILLE, R.T.P. (1970) The pulmonary reaction to high concentrations of fibrous glass dust. *Arch. environ. Health*, **20**: 696-704.

GROSS, P., HARLEY, R.A., & DAVIS, J.M.G. (1976) The lungs of fibre glass workers: comparison with the lungs of a control population. In: *Occupational exposure to fibrous glass. Proceedings of a Symposium, College Park, Maryland, 26-27 June 1974*, Washington DC, US Department of Health, Education and Welfare, pp. 249-263.

HALLIN, N. (1981) *Mineral wool dust in construction sites*, Sweden, 36 pp (Bygghalsan Report 1981-09-01).

HAMMAD, Y.Y. (1984) Deposition and elimination of MMMF. In: *Biological Effects of Man-Made Mineral Fibres. Proceedings of a WHO/IARC Conference, Copenhagen, Denmark, 20-22 April 1982*, Copenhagen, World Health Organization, Regional Office for Europe, **Vol. 2**, pp. 126-142.

HAMMAD, Y.Y. & ESMEN, N.A. (1984) Long-term survey of airborne fibres in the United States. In: *Biological Effects of Man-Made Mineral Fibres. Proceedings of a WHO/IARC Conference, Copenhagen, Denmark, 20-22 April 1982*, Copenhagen, World Health Organization, Regional Office for Europe, **Vol. 1**, pp. 118-132.

HAMMAD, Y.Y., DIEM, J., CRAIGHEAD, J., & WEILL, H. (1982) Deposition of inhaled man-made mineral fibres in the lungs of rats. *Ann. occup. Hyg.*, **26**: 179-187.

HARVEY, G., PAGE, M., & DUMAS, L. (1984) Binding of environmental carcinogens to asbestos and mineral fibres. *Br. J. ind. Med.*, **41**: 396-400.

HAUGEN, A., SCHAFER, P.W., LECHNER, J.F., STONER, G.D., TRUMP, B.F., & HARRIS, C.C. (1982) Cellular ingestion, toxic effects, and lesions observed in human bronchial epithelial tissue and cells cultured with asbestos and glass fibers. *Int. J. Cancer*, **30**: 265-272.

HEAD, I.W.H. & WAGG, R.M. (1980) A survey of occupational exposure to man-made mineral fibre dust. *Ann. occup. Hyg.*, **23**: 235-258.

HESTERBERG, T.W. & BARRETT, J.C. (1984) Dependence of asbestos- and mineral dust-induced transformation of mammalian cells in culture on fiber dimension. *Cancer Res.*, **44**: 2170-2180.

HESTERBERG, T.W., BUTTERICK, C.J., OSHIMURA, M., BRODY, A.R., & BARRETT, J.C. (1986) Role of phagocytosis in Syrian hamster cell transformation and cytogenetic effects induced by asbestos and short and long glass fibres. *Cancer Res.*, **46**: 5795-5802.

HILL, J.W. (1977) Health aspects of man-made mineral fibres. A review. *Ann. occup. Hyg.*, **20**: 161-173.

HILL, J.W. (1978) Man-made mineral fibres. *J. Soc. Occup. Med.*, **28**: 134-141.

HILL, J.W., WHITEHEAD, W.S., CAMERON, J.D., & HEDGECOCK, G.A. (1973) Glass fibres: absence of pulmonary hazard in production workers. *Br. J. ind. Med.*, **30**: 174-179.

HILL, J.W., ROSSITER, C.E., & FODEN, D.W. (1984) A pilot respiratory morbidity study of workers in a MMMF plant in the United Kingdom. In: *Biological Effects of Man-Made Mineral Fibres, Proceedings of a WHO/IARC Conference, Copenhagen, Denmark, 20-22 April 1982*, Copenhagen, World Health Organization, Regional Office for Europe, **Vol. 1**, pp. 413-426.

HMSO (1899) *Her Majesty's Inspector of Factories and Workshops: annual report for 1899*, London, Her Majesty's Stationery Office.

HMSO (1911) *Her Majesty's Inspector of Factories and Workshops: annual report for 1911*, London, Her Majesty's Stationery Office.

HOHR, D. (1985) [Investigations by means of transmission electron microscopy (TEM). Fibrous particles in ambient air.] *Staub-Reinhalt. Luft*, **45**: 171-174 (in German with English summary).

HOLMES, A., MORGAN, A., & DAVISON, W. (1983) Formation of pseudo-asbestos bodies on sized glass fibres in the hamster lung. *Ann. occup. Hyg.*, **27**: 301-313.

HOWIE, R.M., ADDISON, J., CHERRIE, J., ROBERTSON, A., & DODGSON, J. (1986) Letter to the editor. Fibre release from filtering facepiece respirators. *Ann. occup. Hyg.*, **30**: 131-133.

HSC (1979) *Man-made mineral fibres*, London, Health and Safety Commission, 36 pp (Report of a Working Party to the Advisory Committee on Toxic Substances).

IARC (1985) *Final report. Historical cohort study on man-made mineral fibre (MMMF) production workers in seven European countries: extension of the follow-up until 1982*, Lyons, International Agency for Research on Cancer (Report to the Joint European Medical Research Board).

IARC (in press) *Man-made fibres, mineral fibres and radon*, Lyons, International Agency for Research on Cancer (Monographs on the Evaluation of the Carcinogenic Risk of Chemicals to Humans).

ILO (1980) *Guidelines for the use of the ILO international classification of radiographs of pneumoconioses*, revised ed., Geneva, International Labour Organisation (Occupational Safety and Health Series 22).

INDULSKI, J., GOSCICKI, J., WIECEK, E., & STROSZEJN-MROWCA, G. (1984) The evaluation of occupational exposure of workers to airborne MMMF in Poland. In: *Biological Effects of Man-Made*

Mineral Fibres, Proceedings of a WHO/IARC Conference, Copenhagen, Denmark, 20-22 April 1982, Copenhagen, World Health Organization, Regional Office for Europe, **Vol. 1**, pp. 191-197.

JARVHOLM, B. (1984) WHO/IARC meeting on biological effects of man-made mineral fibres. *Eur. J. respir. Dis.*, **65**: 315-316.

JAURAND, M.C., MAGNE, L., BIGNON, J., & GONI, J. (1980) Effects of well-defined fibres on red blood cells and alveolar macrophages. In: Wagner, J.C., ed. *Biological effects of mineral fibres*, Lyons, International Agency for Research on Cancer, **Vol. 1**, pp. 441-450 (IARC Scientific Publication 30).

JOHNSON, N.F. & WAGNER, J.C. (1980) A study by electron microscopy of the effects of chrysotile and man-made mineral fibres on rat lungs. In: Wagner, J.C., ed. *Biological effects of mineral fibres*, Lyons, International Agency for Research on Cancer, **Vol. 1**, pp. 293-303 (IARC Scientific Publication 30).

JOHNSON, N.F., GRIFFITHS, D.M., & HILL, R.J. (1984a) Size distribution following long-term inhalation of MMMF. In: *Biological Effects of Man-Made Mineral Fibres. Proceedings of a WHO/IARC Conference, Copenhagen, Denmark, 20-22 April 1982*, Copenhagen, World Health Organization, Regional Office for Europe, **Vol. 2**, pp. 102-125.

JOHNSON, N.F., LINCOLN, J.L., & WILLS, H.A. (1984b) Analysis of fibres recovered from lung tissue. *Lung*, **162**: 37-47.

KHORAMI, J., LEMIEUX, A., DUNNIGAN, J., & NADEAU, D. (1986) Induced conversion of aluminium silicate fibers into mullite and cristobalite by elevated temperatures: a comparative study on two commercial products. In: *Proceedings of the 15th North American Thermal Analysis Society Conference, Cincinnati, Ohio, 21-24 September 1986*, pp. 343-350 (Paper No. 74).

KILBURN, K.H. (1982) Flame-attenuated fibreglass: another asbestos? *Am. J. ind. Med.*, **3**: 121-125.

KIRK-OTHMER (1980) *Encyclopedia of chemical technology*, New York, John Wiley and Sons.

KLINGHOLZ, R. (1977) Technology and production of man-made mineral fibres. *Ann. occup. Hyg.*, **20**: 153-159.

KLINGHOLZ, R. & STEINKOPF, B. (1984) The reactions of MMMF in a physiological model fluid and in water. In: *Biological Effects of Man-Made Mineral Fibres. Proceedings of a WHO/IARC Conference, Copenhagen, Denmark, 20-22 April 1982*, Copenhagen,

World Health Organization, Regional Office for Europe, **Vol. 2**, pp. 60-86.

KONZEN, J.L. (1976) Results of environmental air-sampling studies conducted in Owens-Corning fiberglass manufacturing plants. In: *Occupational Exposure to Fibrous Glass. Proceedings of a Symposium, College Park, Maryland, 26-27 June 1974*, Washington DC, US Department of Health, Education and Welfare, pp. 115-129.

KROWKE, R., BLUTH, U., MERKER, H.-J., & NEUBERT, D. (1985) Placental transfer and possible teratogenic potential of asbestos in mice. *Teratology*, **32**: 26A-27A.

KUSCHNER, M. (in press) A review of experimental studies on the effects of MMMF on animal systems. *Ann. occup. Hyg.* **31**(4B).

KUSCHNER, M. & WRIGHT, G.W. (1976) The effects of intratracheal instillation of glass fiber of varying size in guinea-pigs. In: *Occupational Exposure to Fibrous Glass. Proceedings of a Symposium, College Park, Maryland, 26-27 June 1974*, Washington DC, US Department of Health, Education and Welfare, pp. 151-168.

LAFUMA, J., MORIN, M., PONCY, J.L., MASSE, R., HIRSCH, A., BIGNON, J., & MONCHAUX, G. (1980) Mesothelioma induced by intrapleural injection of different types of fibers in rats; synergistic effect of other carcinogens. In: Wagner, J.C., ed. *Biological effects of mineral fibres*, Lyons, International Agency for Research on Cancer, **Vol. 1**, pp. 311-320 (IARC Scientific Publication 30).

LE BOUFFANT, L., HENIN, J.P., MARTIN, J.C., NORMAND, C., TICHOUX, G., & TROLARD, F. (1984) Distribution of inhaled MMMF in the rat lung: long-term effects. In: *Biological Effects of Man-Made Mineral Fibres. Proceedings of a WHO/IARC Conference, Copenhagen, Denmark, 20-22 April 1982*, Copenhagen, World Health Organization, Regional Office for Europe, **Vol. 2**, pp. 143-168.

LE BOUFFANT, L., DANIEL, H., HENIN, J.P., MARTIN, J.C., NORMAND, C., TICHOUX, G., & TROLARD, F. (in press) Experimental study on long-term effects of MMMF on the lung of rats. *Ann. occup. Hyg.* **31**(4B).

LEE, K.P., BARRAS, C.E., GRIFFITH, F.D., & WARITZ, R.S. (1979) Pulmonary response to glass fiber by inhalation exposure. *Lab. Invest.*, **40**: 123-133.

LEE, K.P., BARRAS, C.E., GRIFFITH, F.D., WARITZ, R.S., & LAPIN, C.A. (1981) Comparative pulmonary responses to inhaled inorganic fibers with asbestos and fiberglass. *Environ. Res.*, **24**: 167-191.

LEINEWEBER, J.P. (1984) Solubility of fibres *in vitro* and *in vivo*. In: *Biological Effects of Man-Made Mineral Fibres. Proceedings of a WHO/IARC Conference, Copenhagen, Denmark, 20-22 April 1982*, Copenhagen, World Health Organization, Regional Office for Europe, **Vol. 2**, pp. 87-101.

LIPKIN, L.E. (1980) Cellular effects of asbestos and other fibers: correlations with *in vivo* induction of pleural sarcoma. *Environ. Health Perspect.*, **34**: 91-102.

LOCKEY, J. (in press) Respiratory morbidity of man-made vitreous fibre production workers: a prospective study. *Ann. occup. Hyg.* **31**(4B).

LUCAS, J.B (1976) The cutaneous and ocular effects resulting from worker exposure to fibrous glass. In: *Occupational Exposure to Fibrous Glass. Proceedings of a Symposium, College Park, Maryland, 26-27 June 1974*, Washington DC, US Department of Health, Education and Welfare, pp. 211-215.

MCCONNELL, E.E., BASSON, P.A., DEVOS, V., MYERS, B.J., & KUNTZ, R.E. (1974) A survey of disease among 100 free-ranging baboons (*Papio ursenies*) from the Kruger National Park. *Onderstepoort J. vet. Res.*, **41**: 97-168.

MCCONNELL, E.E., WAGNER, J.C., SKIDMORE, J.W., & MOORE, J.A. (1984) A comparative study of the fibrogenic and carcinogenic effects of UICC Canadian chrysotile B asbestos and glass microfibre (JM 100). In: *Biological Effects of Man-Made Mineral Fibres. Proceedings of a WHO/IARC Conference, Copenhagen, Denmark, 20-22 April 1982*, Copenhagen, World Health Organization, Regional Office for Europe, **Vol. 2**, pp. 234-252.

MAGGIONI, A., MEREGALLI, G., SALA, C., & RIVA, M. (1980) [Respiratory and skin diseases in glass fibre workers.] *Med. Lav.*, **71**: 216-227 (in Italian with English abstract).

MALMBERG, P., HEDENSTROM, H., KOLMODIN-HEDMAN, B., & KRANTZ, S. (1984) Pulmonary function in workers of a mineral rock fibre plant. In: *Biological Effects of Man-Made Mineral Fibres. Proceedings of a WHO/IARC Conference, Copenhagen, Denmark, 20-22 April 1982*, Copenhagen, World Health Organization, Regional Office for Europe, **Vol. 1**, pp. 427-435.

MARCONI, A., CORRADETTI, E., & MANNOZZI, A. (in press) Concentrations of man-made vitreous fibres during installation of insulation materials aboard ships at Ancona Naval dockyards. *Ann. occup. Hyg.* **31**(4B).

MAROUDAS, N.G., O'NEILL, C.H., & STANTON, M.F. (1973) Fibroblast anchorage in carcinogenesis by fibres. *Lancet,* **1**: 807-809.

MARSH J.P., JEAN, L., & MOSSMAN, B.T. (1975) Asbestos and fibrous glass induce biosynthesis of polyamines in tracheobronchial epithelial cells *in vitro*. In: Beck, E.G. & Bignon, J., ed. *In vitro effects of mineral dusts*, Berlin, Heidelberg, Springer-Verlag (NATO ASI Series Vol. 63).

MILLER, K. (1980) The *in vivo* effects of glass fibres on alveolar macrophage membrane characteristics. In: Wagner, J.C., ed. *Biological effects of mineral fibres*, Lyons, International Agency for Research on Cancer, **Vol. 1**, pp. 459-465 (IARC Scientific Publication 30).

MITCHELL, R.I., DONOFRIO, D.J., & MOORMAN, W.J. (1986) Chronic inhalation toxicity of fibrous glass in rats and monkeys. *J. Am. Coll. Toxicol.*, **5**(6): 545-575.

MOHR, J.G. & ROWE, W.P. (1978) Fiber glass blown wool or insulation products and their application. In: *Fiber glass*, New York, Van Nostrand Reinhold, pp. 17-189.

MOHR, U., POTT, F., & VONNAHME, F.J. (1984) Morphological aspects of mesotheliomas after intratracheal instillations of fibrous dusts in Syrian golden hamsters. *Exp. Pathol.*, **26**: 179-183.

MONCHAUX, G., BIGNON, J., JAURAND, M.C., LAFUMA, J., SEBASTIEN, P., MASSE, R., HIRSCH, A., & GONI, J. (1981) Mesotheliomas in rats following inoculation with acid-leached chrysotile asbestos and other mineral fibres. *Carcinogenesis*, **2**: 229-236.

MONCHAUX, G., BIGNON, J., HIRSCH, A., & SEBASTIEN, P. (1982) Translocation of mineral fibres through the respiratory system after injection into the pleural cavity of rats. *Ann. occup. Hyg.*, **26**: 309-318.

MORGAN, A. & HOLMES, A. (1984a) Solubility of rockwool fibres *in vivo* and the formation of pseudo-asbestos bodies. *Ann. occup. Hyg.*, **28**: 307-314.

MORGAN, A. & HOLMES, A. (1984b) The deposition of MMMF in the respiratory tract of the rat, their subsequent clearance, solubility *in vivo* and protein coating. In: *Biological Effects of Man-Made Mineral Fibres. Proceedings of a WHO/IARC Conference, Copenhagen, Denmark, 20-22 April 1982*, Copenhagen,

World Health Organization, Regional Office for Europe, **Vol. 2**, pp. 1-17.

MORGAN, A. & HOLMES, A. (1986) Solubility of asbestos and man-made mineral fibers *in vitro* and *in vivo*: its significance in lung disease. *Environ. Res.*, **39**: 475-484.

MORGAN, A., BLACK, A., EVANS, N., HOLMES, A., & PRITCHARD, J.N. (1980) Deposition of sized glass fibres in the respiratory tract of the rat. *Ann. occup. Hyg.*, **23**: 353-366.

MORGAN, A., HOLMES, A., & DAVISON, W. (1982) Clearance of sized glass fibres from the rat lung and their solubility *in vivo*. *Ann. occup. Hyg.*, **25**: 317-331.

MORGAN, R.W., KAPLAN, S.D., & BRATSBERG, J.A. (1984) Mortality in fibrous glass production workers. In: *Biological Effects of Man-Made Mineral Fibres. Proceedings of a WHO/IARC Conference, Copenhagen, Denmark, 20-22 April 1982*, Copenhagen, World Health Organization, Regional Office for Europe, **Vol. 1**, pp. 340-346.

MORRISET, Y., P'AN, A., & JEGIER, Z. (1979) Effect of styrene and fiberglass on small airways of mice. *J. Toxicol. environ. Health*, **5**: 943-956.

MORRISON, D.G., DANIEL, J., LYND, F.T., MOYER, M.P., ESPARZA, R.J., MOYER, R.C., & ROGERS, W. (1981) Retinyl palmitate and ascorbic acid inhibit pulmonary neoplasms in mice exposed to fiberglass dust. *Nutr. Cancer*, **3**: 81-85.

MOULIN, J.J., MUR, J.M., WILD, P., PERREAUX, J.P., & PHAM, Q.T. (1986) Oral cavity and laryngeal cancers among man-made mineral fiber production workers. *Scand. J. Work environ. Health*, **12**: 27-31.

MOULIN, J.J., PHAM, Q.T., MUR, J.M., MEYER-BISCH, C., CAILLARD, J.F., MASSIN, N., WILD, P., TECULESCU, D., DELEPINE, P., HUNZINGER, E., PERREAUX, J.P., MULLER, J., BETZ, M., BAUDIN, V., FONTANA, J.M., HENQUEL, J.C., & TOAMAIN, J.P. (1987) Enquête épidémiologique dans deux usines productrices de fibres minérales artificielles. II. Symptômes respiratoires et fonction pulmonaire. *Arch. Mal. prof. Méd. Trav. Sécur. soc.*, **48**: 7-16.

MUHLE, H., POTT, F., BELLMANN, B., TAKENAKA, S., & ZIEM, U. (in press) Inhalation and injection experiments in rats for testing man-made mineral fibres on carcinogenicity. *Ann. occup. Hyg.* **31**(4B).

NADEAU, D., PARADIS, D., GAUDREAU, A., PELE, J.P., & CALVERT, R. (1983) Biological evaluation of various natural and man-made mineral fibers: cytotoxicity, hemolytic activity and chemiluminescence study. *Environ. Health Perspect.*, **51**: 374 (Abstract).

NAKATANI, Y. (1983) [Biological effects of mineral fibers on lymphocytes *in vitro.*] *Jpn. J. ind. Health*, **25**: 375-386 (in Japanese with English abstract).

NASR, A.N.M., DITCHEK, T., & SCHOLTENS, P.A. (1971) The prevalence of radiographic abnormalities in the chests of fiber glass workers. *J. occup. Med.*, **13**: 371-376.

NEWBALL, H.H. & BRAHIM, S.A. (1976) Respiratory response to domestic fibrous glass exposure. *Environ. Res.*, **12**: 201-207.

NIELSEN, O. (1987) Man-made mineral fibres in the indoor climate caused by ceilings of man-made mineral wool. In: Seifert, B., Esdorn, H., Fischer, M., Rüden, H., & Wegner, J., ed. *Proceedings of the 4th International Conference on Indoor Air Quality and Climate, Berlin (West), 17-21 August 1987*, Berlin, Institute for Water, Soil, and Air Hygiene, **Vol. 1**, pp. 580-583.

NIOSH (1977) *Criteria for a recommended standard. Occupational exposure to fibrous glass*, Cincinnati, Ohio, National Institute for Occupational Safety and Health (DHEW Publication No. 77-152).

NRC (1984) *Asbestiform fibers. Nonoccupational health risks*, Washington DC, National Research Council, National Academy Press, 334 pp.

OHBERG, I. (in press) Technological development of the mineral wool industry in Europe. *Ann. occup. Hyg.* **31**(4B).

OLSEN, J.H., JENSEN, O.M., & KAMPSTRUP, O. (1986) Influence of smoking habits and place of residence on the risk of lung cancer among workers in one rock-wool producing plant in Denmark. *Scand. J. Work environ. Health*, **12**(Suppl. 1): 48-52.

OSHIMURA, M., HESTERBERG, T.W., TSUTSUI, T., & BARRET, J.C. (1984) Correlation of asbestos-induced cytogenetic effects with cell transformation of Syrian hamster embryo cells in culture. *Cancer Res.*, **44**: 5017-5022.

OTTERY, J., CHERRIE, J.W., DODGSON, J., & HARRISON, G.E. (1984) A summary report on environmental conditions at 13 European MMMF plants. In: *Biological Effects of Man-Made Mineral Fibres. Proceedings of a WHO/IARC Conference, Copenhagen, Denmark, 20-22*

April 1982, Copenhagen, World Health Organization, Regional Office for Europe, **Vol. 1**, pp. 83-117.

OTTOLENGHI, A.C., JOSEPH, L.B., NEWMAN, H.A.I., & STEPHENS, R.E. (1983) Interaction of erythrocyte membranes with particulates. *Environ. Health Perspect.*, **51**: 253-256.

PICKRELL, J.A., HILL, J.O., CARPENTER, R.L., HAHN, F.F., & REBAR, A.H. (1983) *In vitro* and *in vivo* response after exposure to man-made mineral and asbestos insulation fibers. *Am. Ind. Hyg. Assoc. J.*, **44**: 557-561.

PIGOTT, G.H. & ISHMAEL, J. (1981) An assessment of the fibrogenic potential of two refractory fibres by intraperitoneal injection in rats. *Toxicol. Lett.*, **8**: 153-163.

PIGOTT, G.H., GASKELL, B.A., & ISHMAEL, J. (1981) Effects of long-term inhalation of alumina fibres in rats. *Br. J. exp. Pathol.*, **62**: 323-331.

POSSICK, P.A., GELLIN, G.A., & KEY, M.M. (1970) Fibrous glass dermatitis. *Am. ind. Hyg. J.*, **31**: 12-15.

POTT, F. (1978) Some aspects on the dosimetry of the carcinogenic potency of asbestos and other dusts. *Staub-Reinhalt. Luft*, **38**: 486.

POTT, F., HUTH, F., & FRIEDRICHS, K.H. (1974) Tumorigenic effect of fibrous dusts in experimental animals. *Environ. Health Perspect.*, **9**: 313-315.

POTT, F., HUTH, F., & SPURNY, K. (1980) Tumour induction after intraperitoneal injection of fibrous dusts. In: Wagner, J.C., ed. *Biological effects of mineral fibres*, Lyons, International Agency for Research on Cancer, **Vol. 1**, pp. 337-342 (IARC Scientific Publication 30).

POTT, F., SCHLIPKOTER, H.W., ZIEM, U., SPURNY, K., & HUTH, F. (1984) New results from implantation experiments with mineral fibres. In: *Biological Effects of Man-Made Mineral Fibres. Proceedings of a WHO/IARC Conference, Copenhagen, Denmark, 20-22 April 1982*, Copenhagen, World Health Organization, Regional Office for Europe, **Vol. 2**, pp. 286-302.

POTT, F., ZIEM, U., REIFFER, F.J., HUTH, F., ERNST, H., & MOHR, U. (1987) Carcinogenicity studies in fibres, metal compounds, and some other dusts in rats. *Exp. Pathol.*, **32**: 129-152.

POTT, F., ROLLER, M., ZIEM, U., REIFFER, F.-J., BELLMANN, B., ROSENBRUCH, M., & HUTH, F. (in press) Carcinogenicity studies on natural and man-made fibres with the intraperitoneal tests in rats. In: *Proceedings of the Symposium on Mineral Fibres in the Non-Occupational Environment, Lyons, 8-10 September 1987*, Lyons, International Agency for Research on Cancer.

PYLEV, L.N., KOVALSKAYA, G.D., & YAKOVENKO, G.N. (1975) [Carcinogenic activity of synthetic asbestos.] *Gig. Tr. prof. Zabol.*, **10**:31-39.

RENDALL, R.E.G. & SCHOEMAN, J.J. (1985) A membrane filter technique for glass fibres. *Ann. occup. Hyg.*, **29**: 101-108.

RENNE, R.A., ELDRIDGE, S.R., LEWIS, T.R., & STEVENS, D.L. (1985) Fibrogenic potential of intratracheally instilled quartz, ferric oxide, fibrous glass, and hydrated alumina in hamsters. *Toxicol. Pathol.*, **13**: 306-314.

REUZEL, P.G.J., FERON, V.J., SPIT, B.J., BEEMS, R.B., & KROES, R. (1983) Tissue damage and nutritional factors in experimental respiratory tract (co-)carcinogenesis. *Environ. Health Perspect.*, **50**: 275-283.

RICHARDS, R.J. & JACOBY, F. (1976) Light microscope studies on the effects of chrysotile asbestos and fiber glass on the morphology and reticulin formation of cultured lung fibroblasts. *Environ. Res.*, **11**: 112-121.

RICHARDS, R.J. & MORRIS, F. (1973) Collagen and mucopolysaccharide production in growing lung fibroblasts exposed to chrysotile asbestos. *Life Sci.*, **12**: 441-451.

RIEDIGER, G. (1984) Measurements of mineral fibres in the industries which produce and use MMMF. In: *Biological Effects of Man-Made Mineral Fibres. Proceedings of a WHO/IARC Conference, Copenhagen, Denmark, 20-22 April 1982*, Copenhagen, World Health Organization, Regional Office for Europe, **Vol. 1**, pp. 133-177.

RINDEL, A., BACH, E., BREUM, N.O., HUGOD, C., & SCHNEIDER, T. (1987) Correlating health effect with indoor air quality in kindergartens. *Int. Arch. occup. environ. Health*, **59**: 363-373.

RIRIE, D.G., HESTERBERG, T.W., BARRETT, J.C., & NETTESHEIM, P. (1985) Toxicity of asbestos and glass fibers for rat tracheal epithelial cells in culture. In: Beck, E.G. & Bignon, J., ed. *In vitro effects of mineral dusts*, Berlin, Heidelberg, Springer-Verlag, pp. 177-184 (NATO ASI Series, **Vol. G3**).

ROBINSON, C.F., DEMENT, J.M., NESS, G.O., & WAXWEILER, R.J. (1982) Mortality patterns of rock and slag mineral wool production workers: an epidemiological and environmental study. *Br. J. ind. Med.*, **39**: 45-53.

ROOD, A.P. & STREETER, R.R. (1985) Size distributions of airborne superfine man-made mineral fibers determined by transmission electron microscopy. *Am. Ind. Hyg. Assoc. J.*, **46**: 257-261.

ROSCHIN, A.V. & AZOVA, S.M. (1975) [Dust factor in production of new types of fibrous glass.] *Gig. i Sanit.*, **12**: 24-28 (in Russian).

ROWHANI, F. & HAMMAD, Y.Y. (1984) Lobar deposition of fibers in the rat. *Am. Ind. Hyg. Assoc. J.*, **45**: 436-439.

SARACCI, R. (1980) Introduction: epidemiology of groups exposed to other mineral fibres. In: Wagner, J.C., ed. *Biological effects of mineral fibres*, Lyons, International Agency for Research on Cancer, **Vol. 2**, pp. 951-963 (IARC Scientific Publication 30).

SARACCI, R. & SIMONATO, L. (1982) Man-made vitreous fibers and workers' health. *Scand. J. Work environ. Health*, **8**: 234-242.

SARACCI, R., SIMONATO, L., ACHESON, E.D., ANDERSEN, A., BERTAZZI, P.A., CLAUDE, J., CHARNAY, N., ESTEVE, J., FRENTZEL-BEYME, R.R., GARDNER, M.J, JENSEN, O.M., MAASING, R., OLSEN, J.H., TEPPO, L., WESTERHOLM, P., & ZOCCHETTI, C. (1984a) Mortality and incidence of cancer of workers in the man made vitreous fibres producing industry: an international investigation at 13 European plants. *Br. J. ind. Med.*, **41**: 425-436.

SARACCI, R., SIMONATO, L., ACHESON, E.D., ANDERSEN, A., BERTAZZI, P.A., CLAUDE, J., CHARNAY, N., ESTEVE, J., FRENTZEL-BEYME, R.R., GARDNER, M.J., JENSEN, O.M., MAASING, R., OLSEN, J.H., TEPPO, L.H.I., WESTERHOLM, P., & ZOCCHETTI, C. (1984b) The IARC mortality and cancer incidence study of MMMF production workers. In: *Biological Effects of Man-Made Mineral Fibres. Proceedings of a WHO/IARC Conference, Copenhagen, Denmark, 20-22 April 1982*, Copenhagen, World Health Organization, Regional Office for Europe, **Vol. 1**, pp. 279-310.

SCHEPERS, G.W.H. (1955) The biological action of glass wool: studies on experimental pulmonary histopathology. *Am. Med. Assoc. Arch. Ind. Health*, **12**: 280.

SCHEPERS, G.W.H. & DELAHUNT, A.B. (1955) An experimental study of the effects of glass wool on animal lungs. *Am. Med. Assoc. Arch. Ind. Health*, **12**: 276-279.

SCHNEIDER, T. (1979) Exposures to man-made mineral fibres in user industries in Scandinavia. *Ann. occup. Hyg.*, **22**: 153-162.

SCHNEIDER, T. (1984) Review of surveys in industries that use MMMF. In: *Biological Effects of Man-Made Mineral Fibres. Proceedings of a WHO/IARC Conference, Copenhagen, Denmark, 20-22 April 1982*, Copenhagen, World Health Organization, Regional Office for Europe, **Vol. 1**, pp. 178-190.

SCHNEIDER, T. (1986) Manmade mineral fibers and other fibers in the air and in settled dust. *Environ. Int.*, **12**: 61-65.

SCHNEIDER, T. & STOKHOLM, J. (1981) Accumulation of fibers in the eyes of workers handling man-made mineral fiber products. *Scand. J. Work environ. Health*, **7**: 271-276.

SCHOLZE, H. & CONRADT, R. (in press) *In vitro* study on siliceous fibres. *Ann. occup. Hyg.* **31**(4B).

SHANNON, H.S., HAYES, M., JULIAN, J.A., & MUIR, D.C.F. (1984a) Mortality experience of glass fibre workers. In: *Biological Effects of Man-Made Mineral Fibres. Proceedings of a WHO/IARC Conference, Copenhagen, Denmark, 20-22 April 1982*, Copenhagen, World Health Organization, Regional Office for Europe, **Vol. 1**, pp. 347-349.

SHANNON, H.S., HAYES, M., JULIAN, J.A., & MUIR, D.C.F. (1984b) Mortality experience of glass fibre workers. *Br. J. ind. Med.*, **41**: 35-38.

SHANNON, H.S., JAMIESON, E., JULIAN, J.A., MUIR, D.C.F., & WALSH, C. (in press) Mortality experience of glass fibre workers: extended follow-up. *Ann. occup. Hyg.* **31**(4B).

SIMONATO, L., FLETCHER, A.C., CHERRIE, J., ANDERSEN, A., BERTAZZI, P., CHARNAY, N., CLAUDE, J., DODGSON, J., ESTEVE, J., FRENTZEL-BEYME, R., GARDNER, M.J., JENSEN, O., OLSEN, J., SARACCI, R., TEPPO, L., WINKELMANN, R., WESTERHOLM, P., WINTER, P.D., & ZOCCHETTI, C. (1986a) The man-made mineral fiber European historical cohort study: extension of the follow-up. *Scand. J. Work environ. Health*, **12**(Suppl. 1): 34-47.

SIMONATO, L., FLETCHER, A.C., CHERRIE, J., ANDERSEN, A., BERTAZZI, P., CHARNAY, N., CLAUDE, J., DODGSON, J., ESTEVE, J., FRENTZEL-BEYME, R., GARDNER, M.J., JENSEN, O., OLSEN, J., SARACCI, R., TEPPO, L., WINKELMANN, R., WESTERHOLM, P., WINTER, P.D., & ZOCCHETTI, C. (1986b) Updating lung cancer mortality among a cohort of man-made mineral fibre production workers in seven European countries. *Cancer Lett.*, **30**: 189-200.

SIMONATO, L., FLETCHER, A.C., CHERRIE, J., ANDERSEN, A., BERTAZZI, P., CHARNAY, N., CLAUDE, J., DODGSON, J., ESTEVE, J., FRENTZEL-BEYME, R., GARDNER, M.J., JENSEN, O., OLSEN, J., SARACCI, R., TEPPO, L., WINKELMANN, R., WESTERHOLM, P., WINTER,

P.D., & ZOCCHETTI, C. (in press) The man-made mineral fibres (MMMF) European historical cohort study: extension of the follow-up. *Ann. occup. Hyg.* **31**(4B).

SINCOCK, A. & SEABRIGHT, M. (1975) Induction of chromosome changes in Chinese hamster cells by exposure to asbestos fibres. *Nature (Lond.)*, **257**(5521): 56-58.

SINCOCK, A.M., DELHANTY, J.D., & CASEY, G. (1982) A comparison of the cytogenetic response to asbestos and glass fibre in Chinese hamster and human cell lines. Demonstration of growth inhibition in primary human fibroblasts. *Mutat. Res.*, **101**: 257-268.

SIXT, R., BAKE, B., ABRAHAMSSON, G., & THIRINGER, G. (1983) Lung function of sheet metal workers exposed to fiber glass. *Scand. J. Work environ. Health*, **9**: 9-14.

SKOMAROKHIN, A.F. (1985) [*Dust factors at production and application of new types of man-made mineral fibres: its effects on organisms*,] Leningrad, Sanitary Hygienic Medical Institution (Thesis) (in Russian).

SMITH, D.M., ORTIZ, L.W., & ARCHULETA, R.F. (1984) Long-term exposure of Syrian hamsters and Osborne-Mendel rats to aerosolized 0.45 μm mean diameter fibrous glass. In: *Biological Effects of Man-Made Mineral Fibres. Proceedings of a WHO/IARC Conference, Copenhagen, Denmark, 20-22 April 1982*, Copenhagen, World Health Organization, Regional Office for Europe, **Vol.** 2, pp. 253-272.

SMITH, D.M., ORTIZ, L.W., ARCHULETA, R.F., & JOHNSON, N.F. (in press) Long-term health effects in hamsters and rats exposed chronically to man-made vitreous fibers. *Hyg. occup. Hyg.* **31**(4B).

SMITH, W.E., HUBERT, D.D., & SOBEL, H.J. (1980) Dimensions of fibres in relation to biological activity. In: Wagner, J.C., ed. *Biological effects of mineral fibres*, Lyons, International Agency for Research on Cancer, **Vol.** 1, pp. 357-360 (IARC Scientific Publication 30).

SNIPES, M.B., YEH, H.C., OLSON, T.R., & NARVAIZ, R.J. (1984) Deposition and retention patterns for 3, 9, and 15 μm latex microspheres inhaled by rats and guinea-pigs. In: *The Annual Report 1983-84 of the Inhalation Toxicology Research Institute*, Albuquerque, New Mexico.

SPURNY, K.R. (1983) Measurement and analysis of chemically changed mineral fibers after experiments *in vitro* and *in vivo*. *Environ. Health Perspect.*, **51**: 343-355.

SPURNY, K.R., POTT, F., STOBER, W., OPIELA, H., SCHORMANN, J., & WEISS, G. (1983) On the chemical changes of asbestos fibers and MMMFs in biologic residence and in the environment. Part 1. *Am. Ind. Hyg. Assoc. J.*, **44**: 833-845.

STAHULJAK-BERITIC, D., SKURIC, Z., VALIC, F., & MARK, B. (1982) Respiratory symptoms, ventilatory function, and lung X-ray changes in rock wool workers. *Acta med. Jug.*, **36**: 333-342.

STANTON, M.F. & LAYARD, M. (1978) The carcinogenicity of fibrous minerals. In: *Proceedings of the Workshop on Asbestos: Definitions and Measurement Methods, Gaithersburg, Maryland, 18-20 July 1977*, Gaithersburg, Maryland, National Bureau of Standards, pp. 143-151.

STANTON, M.F. & WRENCH, C. (1972) Mechanisms of mesothelioma induction with asbestos and fibrous glass. *J. Natl Cancer Inst.*, **48**: 797-822.

STANTON, M.F., LAYARD, M., TEGERIS, A., MILLER, E., MAY, M., & KENT, E. (1977) Carcinogenicity of fibrous glass: pleural response in the rat in relation to fiber dimension. *J. Natl Cancer Inst.*, **58**: 587-603.

STANTON, M.F., LAYARD, M., TEGERIS, A., MILLER, E., MAY, M., MORGAN, E., & SMITH, A. (1981) Relation of particle dimension to carcinogenicity in amphibole asbestoses and other fibrous minerals. *J. Natl Cancer Inst.*, **67**: 965-975.

STOKHOLM, J., NORN, M., & SCHNEIDER, T. (1982) Ophthamologic effects of man-made mineral fibers. *Scand. J. Work environ. Health*, **8**: 185-190.

STRUBEL, G., FRAIJ, B., RODELSPERGER, K., & WOITOWITZ, H.J. (1986) Man-made mineral fibers in the working environment. Letter to the editor. *Am. J. ind. Med.*, **10**: 101-102.

STYLES, J.A. & WILSON, J. (1976) Comparison between *in vitro* toxicity of two novel fibrous mineral dusts and their tissue reactions *in vivo*. *Ann. occup. Hyg.*, **19**: 63-68.

SYKES, S.E., MORGAN, A., MOORES, S.R., DAVISON, W., BECK, J., & HOLMES, A. (1983) The advantages and limitations of an *in vivo* test system for investigating the cytotoxicity and fibrogenicity of fibrous dusts. *Environ. Health Perspect.*, **51**: 267-273.

TEPPO, L. & KOJONEN, E. (1986) Mortality and cancer risk among workers exposed to man-made mineral fibers in Finland. *Scand. J. Work environ. Health,* **12**(Suppl. 1): 61-64.

TIESLER, H. (1982) [Production-related omissions by manufacturing man-made mineral fibres.] In: *Proceedings of the International VDI Colloquium: Fibrous Dusts, Strasbourg, 6 October 1982* (in German).

TILKES, F. & BECK, E.G. (1980) Comparison of length-dependent cytotoxicity of inhalable asbestos and man-made mineral fibres. In: Wagner, J.C., ed. *Biological effects of mineral fibres,* Lyons, International Agency for Research on Cancer, **Vol. 1**, pp. 475-483 (IARC Scientific Publication 30).

TILKES, F. & BECK, E.G. (1983a) Influence of well-defined mineral fibers on proliferating cells. *Environ. Health Perspect.*, **51**: 275-279.

TILKES, F. & BECK, E.G. (1983b) Macrophage functions after exposure to mineral fibers. *Environ. Health Perspect.*, **51**: 67-72.

TIMA (1982) *Man-made vitreous fibers and their uses*, New York, Thermal Insulation Manufacturers Association, 1 pp.

TIMBRELL, V. (1965) The inhalation of fibrous dusts. *Ann. NY Acad. Sci.*, **132**: 255-273.

TIMBRELL, V. (1976) Aerodynamic considerations and other aspects of glass fibre. In: *Occupational Exposure to Fibrous Glass. Proceedings of a Symposium, College Park, Maryland, 26-27 June 1974*, Washington DC, US Department of Health, Education and Welfare, pp. 33-50.

TOFT, P. & MEEK, M.E. (1986) Human exposure to asbestos in the environment. In: *Proceedings of the International Conference on Chemicals in the Environment, Lisbon, Portugal, 1-3 July 1986*, London, Selper Ltd., pp. 492-501.

UPTON, A.C. & FINK, D.J. (1979) Pneumoconiosis and fibrous glass. *Am. Ind. Hyg. Assoc. J.*, **40**: A14-A16.

US EPA (1980) *Air quality criteria for particulate matter and sulfur oxide,* Research Triangle Park, US Environmental

Protection Agency, Environmental Criteria and Assessment Office.

UTIDJIAN, M. & COOPER, W.C. (1976) Human epidemiologic studies with emphasis on chronic pulmonary effects. In: *Occupational Exposure to Fibrous Glass. Proceedings of a Symposium, College Park, Maryland, 26-27 June 1974*, Washington DC, US Department of Health, Education and Welfare, pp. 223-224.

VALIC, F. (1983) *ILO Encyclopaedia of occupational health and safety*, **Vol. 1**, Geneva, International Labour Office.

VERBECK, S.J.A., BUISE-VAN UNNIK, E.M.M., & MALTEN, K.E. (1981) Itching in office workers from glass fibres. *Contact Dermatit., 7: 354.*

VINE, G., YOUNG, J., & NOWELL, I.W. (1983) Health hazards associated with aluminosilicate fibre products. Ann. occup. Hyg., **28**: 356-359.

WAGNER, J.C., BERRY, G., & TIMBRELL, V. (1973) Mesotheliomata in rats after inoculation with asbestos and other materials. *Br. J. Cancer*, **28**: 173-185.

WAGNER, J.C., BERRY, G.B., HILL, R.J., MUNDAY, D.E., & SKIDMORE, J.W. (1984) Animal experiments with MMM(V)F fibres-effects of inhalation and intrapleural inoculation in rats. In: *Biological Effects of Man-Made Mineral Fibres. Proceedings of a WHO/IARC Conference, Copenhagen, Denmark, 20-22 April 1982*, Copenhagen, World Health Organization, Regional Office for Europe, **Vol. 2**, pp. 209-233.

WEILL, H., HUGHES, J.M., HAMMAD, Y.Y., GLINDMEYER, H.W., SHARON, G., & JONES, R.N. (1983) Respiratory health in workers exposed to man-made vitreous fibers. *Am. Rev. respir. Dis.*, **128**: 104-112.

WEILL, H., HUGHES, J.M., HAMMAD, Y.Y., GLINDMEYER, H.W., SHARON, G., & JONES, R.N. (1984) Respiratory health of workers exposed to MMMF. In: *Biological Effects of Man-Made Mineral Fibres. Proceedings of a WHO/IARC Conference, Copenhagen, Denmark, 20-22 April 1982*, Copenhagen, World Health Organization, Regional Office for Europe, **Vol. 1**, pp. 387-412.

WESTERHOLM, P. & BOLANDER, A.-M. (1986) Mortality and cancer incidence in the man-made mineral fiber industry in Sweden. *Scand. J. Work environ. Health*, **12**(Suppl. 1): 78-84.

WHO (1981) *Methods of monitoring and evaluating man-made mineral fibres*, Copenhagen, World Health Organization, Regional Office for Europe, 53 pp (EURO Reports and Studies No. 48).

WHO (1983a) *Biological effects of man-made mineral fibres*, Copenhagen, World Health Organization, Regional Office for Europe, 155 pp (EURO Reports and Studies No. 81).

WHO (1983b) *EHC 27: Guidelines on studies in environmental epidemiology*, Geneva, World Health Organization, 351 pp.

WHO (1984) *Evaluation of exposure to airborne particles in the work environment*, Geneva, World Health Organization (WHO Offset Publication No. 80).

WHO (1985) *Reference method for measuring airborne man-made mineral fibres (MMMF)*, Copenhagen, World Health Organization, Regional Office for Europe (Environmental Health Report No. 4).

WOODWORTH, C.D., MOSSMAN, B.T., & CRAIGHEAD, J.E. (1983) Induction of squamous metaplasia in organ cultures of hamster trachea by naturally occurring and synthetic fibers. *Cancer Res.*, **43**: 4906-4912.

WRIGHT, A., GORMLEY, I.P., & DAVIS, J.M.G. (1986) The *in vitro* cytotoxicity of asbestos fibers. I. P388D1 cells. *Am. J. ind. Med.*, **9**: 371-384.

WRIGHT, G.W. (1968) Airborne fibrous glass particles: chest roentgenograms of persons with prolonged exposure. *Arch. environ. Health*, **16**: 175-181.

WRIGHT, G.W. (1984) Respiratory morbidity of MMMF production workers: a review of previous studies. In: *Biological Effects of Man-Made Mineral Fibres. Proceedings of a WHO/IARC Conference, Copenhagen, Denmark, 20-22 April 1982*, Copenhagen, World Health Organization, Regional Office for Europe, **Vol. 1**, pp. 381-386.

WRIGHT, G.W. & KUSCHNER, M. (1977) The influence of varying lengths of glass and asbestos fibres on tissue response in guinea-pigs. In: Walton, W.H., ed. *Inhaled particles. IV. Proceedings of an International Symposium, Edinburgh, 22-26 September 1975*, Oxford, Pergamon Press, pp. 455-474.

Other titles available in the ENVIRONMENTAL HEALTH CRITERIA series (continued):

44. Mirex
45. Camphechlor
46. Guidelines for the Study of Genetic Effects in Human Populations
47. Summary Report on the Evaluation of Short-term Tests for Carcinogens (Collaborative Study on *In Vitro* Tests)
48. Dimethyl Sulfate
49. Acrylamide
50. Trichloroethylene
51. Guide to Short-term Tests for Detecting Mutagenic and Carcinogenic Chemicals
52. Toluene
53. Asbestos and Other Natural Mineral Fibres
54. Ammonia
55. Ethylene Oxide
56. Propylene Oxide
57. Principles of Toxicokinetic Studies
58. Selenium
59. Principles for Evaluating Health Risks from Chemicals During Infancy and Early Childhood: The Need for a Special Approach
60. Principles and Methods for the Assessment of Neurotoxicity Associated With Exposure to Chemicals
61. Chromium
62. 1,2-Dichloroethane
63. Organophosphorus Insecticides - A General Introduction
64. Carbamate Pesticides - A General Introduction
65. Butanols - Four Isomers
66. Kelevan
67. Tetradifon
68. Hydrazine
69. Magnetic Fields
70. Principles for the Safety Assessment of Food Additives and Contaminants in Food
71. Pentachlorophenol
72. Principles of Studies on Diseases of Suspected Chemical Etiology and Their Prevention
73. Phosphine and Selected Metal Phosphides
74. Diaminotoluenes
75. Toluene Diisocyanates
76. Thiocarbamate Pesticides - A General Introduction